Ruminations of a
Grumpy Shepherd

Ruminations of a Grumpy Shepherd

Random Thoughts from a Life of Raising Sheep

By Richard Regnery

Baa Baa Doo Press

Sturgeon Bay, Wisconsin

Ruminations of a Grumpy Shepherd
©2010 by Richard Regnery. Photographs © Richard Regnery

ISBN 978-0-557-34422-2

This book is dedicated to my wife and partner in sheep. Without her there would be little that functions at Whitefish Bay Farm and without her this written work would never have seen the light of day.

Chapters

An Introduction

These random essays began many years ago at the urging of Lanette, a shepherd friend. We met through a sheep discussion group in the early days of the internet. At the time she was a regular contributor to the <u>Black Sheep Newsletter</u>. It was at her urging that I set down some notes about my education and experiences as a shepherd at Whitefish Bay Farm. Also it was at her urging that I share theses notes with the readers of <u>Black Sheep Newsletter</u>. Lanette and I have never met in person and now, after many years, have lost track of each other. I did, nonetheless write that article and have continued to write additional essays over the years. This collection of essays is based upon those articles. With the benefit of time and experience I have expanded and fine-tuned many of theses essays. I have also added newer pieces, some of which are based upon a blog that we now maintain for the farm under the title: "the Ewe Turn".

I wish to thank Peggy Lundquist, the editor of <u>Black Sheep Newsletter</u>, for permission to re-use these essays, albeit in a slightly altered form. Over the years Peggy and her wonderful co-workers have maintained the utmost patience with me. I am, it must be told, an author who rarely makes a deadline. That they continued to encourage me to write is astonishing in itself.

One must understand that historically for me the assignment of a writing project brings back dark childhood memories of English classes where I rarely found the teacher who could inspire me to write. Ever since my youth, I have had an aversion to assigned writing tasks. With the exception of my seventh grade teacher, Miss Ruth Gordon, my travels through essays and creative writing were dark and grim. Only a due date and the threat of failing the course would put me to the task. Strangely, due to positive experiences I had writing letters a bit later in my life, I find no disincentive to correspondence either with others or, in a sense, with myself. So bear with me here. If the efforts that are to follow sound a bit like letters it is perhaps no coincidence.

Shepherd Education

An introduction is in order. My wife and partner, Gretchen, and I have been raising Corriedale sheep for their wool in northeast Wisconsin since 1990. Prior to that time, we had spent much of our adult lives in careers which provided us with a comfortable material lifestyle but ultimately little personal satisfaction. Our experiences of finding adulthood in the turbulent 1960's had always remained with us. It just took us a while to finally throw off the bindings of our contemporary society and do something we "wanted to do" rather than what we "should do". As a result we opted to become shepherds as we aged through our 40's.

Our current odyssey with sheep began in 1983 with the purchase of an abandoned dairy farm in rural Door County, Wisconsin. Our ultimate goal was to support ourselves by raising sheep, using and selling their fiber through an on-farm gallery/studio and to share these experiences with others by operating a bed and breakfast from our farm home. It took us almost seven years of weekend and vacation labors to recondition our buildings so that all these goals could be achieved. Our gallery, in the old granary, was first to become operational in 1983. In the beginning we sold the paintings and fiber works of others, most notably, Gretchen's parents' (both of whom had become

3

handweavers in their 60's). Later the gallery would come to focus on our own fiber production. In 1988 I quit my job with the Federal Government in Milwaukee, found employment on a local dairy farm and moved to the farm. Gretchen followed in less than a year and worked for the local hospital while we both continued to work on our dream. Not until 1990 did we have our barn and pastures sufficiently repaired and set up to welcome sheep. A year later the house was ready to welcome guests, and we were ready to cut ourselves away from outside employment.

The Education Starts

Our education as shepherds began in earnest with the arrival of our first sheep in the summer of 1990. Prior to that time Gretchen had never worked with sheep. My experience had started with a family pet in California when I was quite young. She was known simply as "Ma sheep", a colored ewe of now unknown breed who was reputed to be nearly twenty years of age. Prior to her life with us, she had been the companion to an elephant that live with a wealthy man from our community. It is hard to tell how that experience influenced her; she was kind, friendly and gentle. When the elephant died she returned to her original flock. It was from that flock that we obtained her. She was close to being a "gummer" (not much in the way of teeth) when we purchased her, but she was pregnant and produced a wonderful lamb the following spring. I did not get my own sheep until I was old enough to be in 4-H club. My initial 4-H project was to raise two market lambs to show at the county fair. The lambs were obtained through the aid of our 4-H club. Mine were two mixed breed ewes purchased in the early spring to be shown in the summer once fattened by my efforts. It took a couple of months for us to realize that one of the two was

4

pregnant. That should have been a future sign that I should have heeded! Things do not always go as planned when it comes to sheep. In retrospect it was obvious that the ewe lambs were much older than represented. A ewe lamb rarely becomes pregnant before seven or eight months of age. Since one of my "lambs" was already pregnant before my purchase, she was also much older than the three or four month age she was purported to be. In any case the one "lamb" was not suitable for showing as a feeder at the fair. My "lamb fattening" project rapidly became a "sheep breeding" project. Miraculously the ewe, Sullivan, did just fine with her pregnancy and produced a wonderful, growthy Shropshire-like ewe lamb in the fall. My other lamb, Gilbert, never got to the fair, but rather she stayed on as Sullivan's companion.

Around this time (late 1950's) my mother became interested in handspinning and working with natural colored fibers. Very few people were interested in raising colored sheep in the 50's and there was a dearth of knowledge about colored sheep genetics. The occasional colored lamb would arrival, unexpectedly, the result of the expression of relatively rare recessive colored genetics found in both white parents. It was known at that time that the Karakul breed of sheep was one of the few color dominant breeds found in North America; using a Karakul ram on a white ewe of most any breed would generally produce a colored offspring. We were lucky enough to know someone with a black Karakul ram. We arranged for Sullivan to visit with that ram after she had raised her first lamb. The next year, she produced a ram lamb, Yogi, of wonderful color and evil disposition. About the time that Yogi was old enough to shear and slaughter my ovine experience was put on hold, as my family moved to Australia for a year and the sheep were sold.

In retrospect I now know that during my introduction to sheep, I missed a wonderful opportunity to learn about colored sheep and their genetics. My parents were family friends of Anne Blinks, who, in the late 1950's, was beginning her pioneering work with colored sheep genetics. At the time I did not realize it but her breed of choice was Corriedale. It is ironic and pure coincidence that over 30 years later Gretchen and I are raising them. I deeply regret that I did not know enough in my youth to learn from Anne Blinks.

Our choice of Corriedales was based upon both research and convenience. We were not in a position to be able to afford a rare sheep breed or other fiber animal. Nor did we feel safe investing a large sum of money in just a few rare animals only to run the risk of loosing some or all of them due to our own inexperience. We were looking for a good size sheep which produced good quality wool that was good for handspinning and use in next to the skin clothing. We also wanted a breed where there were some colored animals already available for sale in the upper Midwest. We figured that we did not have either the time or money to travel across the county to purchase breeding ewes or rams for genetic diversity. Those criteria quickly narrowed the choices for breeds in Wisconsin to Corriedales.

In 1990 Gretchen and I purchased eight white and eleven colored Corriedale ewe lambs, each group from an unrelated flock. We also purchased a 5 year old white Corriedale ram, Monty, whose origin was from a distant third and unrelated flock. By 1997 our flock had grown to include includes five rams, 64 ewes and a wether. Less than seven years later the flock reached it maximum size of somewhere around 130 (of which six or seven were rams). The flock cannot grow any further without a significant building project to expand winter housing. The flock is usually about 65% to 75% colored. Of the white ewes most are intentionally the product of colored/white crosses and therefore carry a strong color factor for breeding. They all remain 100% Corriedale.

Beginning Lessons

Rather than make this essay a treatise about what we do and do not do to succeed as shepherds, let me relate some random experiences which reflect some of our original innocence, but which also resulted in our rapid education.

Our first breeding involved the eight white ewe lambs and the older ram, "Monty". We had read all we could and felt that we were adequately prepared for breeding. Besides the sheep, we had all the equipment we felt we needed, including a nice marking harness for the ram. The harness carries a crayon on the ram's chest so that when the ram mounts a ewe he leaves a mark on her. The mark provides us with a breeding date. If the ewe is not remarked within 17 days (her heat cycle) it tells us that the ewe is most likely pregnant. If she is remarked, the clock is re-set to watch for another mark in about 17 days. Rams are not always friendly or shepherd-tolerant, but Monty was a "pussycat" as rams go, for which we were extremely thankful. When the day came for turning him in with his ewes he was patient and cooperative. He needed to be, as we struggled long and mightily to place him in his harness. We could not get the knack of it! Somehow it just did not seem to fit correctly, despite the fact that we read and re-read the instructions along with examining the diagrams. At last we thought we might have it. So off Monty trotted to his girls. He was a very efficient lad. By the next morning there appeared to be a mark on a couple ewes, but the harness was in total disarray. We struggled repeatedly to get the harness back on. This scenario continued for days. Often we would find the harness completely off the ram. In frustration a number of days later, we happened to look at the catalogue from which the harness had been ordered. Low and behold, a photo showed a different installation than that which was drawn in the instructions. Adjustment was made and the harness stayed on perfectly! LESSON: Do not rely implicitly on anyone's written instruction and sketches, even if they come from a country with centuries of experience raising sheep (in this case Great Britain). As a footnote, by the time the harness was properly installed Monty failed to mark a ewe. In a way we were still not certain if it was "working". We had to wait until February to confirm that he had settled all eight ewe lambs on their first heat cycle, which was an amazing feat for a ram with eight flighty ewe lambs. Since that year, properly installed marking harnesses have been invaluable to us.

Lambing by the book

Our first lambing now seems so distant, but it is still so clear. With a total of only eight pregnant ewes (compared to our later totals of 80 to 90 plus) we hoped that things would go well. We took no chances; we read and re-read everything we had. Then we waited and watched. Monty had done a wonderful job, especially since all of his ewes were inexperienced lambs. Lambing lasted about two weeks. When Abigail was the first ewe to go into labor, we stood aside, with all of our books at the ready. Every step of her labor was textbook perfect, except for the fact that she progressed faster than we could read. The image of her peacefully proceeding with her labor while we frantically thumbed through pages would have probably been hilarious to a bystander; we were too stressed to visualize it at the time. When she produced a healthy ewe lamb and proceeded to mother it properly, the two of us were

8

so proud of ourselves. (Beware of such pride!) We gave Abigail sufficient time to lick the lamb clean, for us to get its navel cord clipped and dipped in iodine and finally to make sure the lamb was nursing. We then placed the two in a private jug. While we were congratulating ourselves profusely, Abigail suddenly dropped a second, healthy lamb, literally right at our feet. In her own quiet way she had slapped both of us back to the reality that we did not know it all. Abigail was with us for eleven years, producing twins and triplets with regularity for nine years. When we let her retire at age nine she had produce 22 lambs. Her first lamb, Cassie, also remained in the flock for many years. She was a constant reminder to be humble when it comes to our extent of knowledge.

Continuing Lessons

After three years of lambing, we began to feel a bit more confident in our abilities and knowledge. We rotationally graze our flock from early spring into late October. The lambs remain with their dams all spring and summer. We have always castrated most of the ram lambs before they were a week old, keeping a few intact for sale, or to breed our ewe lambs in the fall and/or to be replacements for our older rams as they were retired. All of our books from that time indicated that we need not be concerned about a ram lamb breeding ewes until he was at least 6 or more months of age. We never experienced a problem leaving these few intact rams lambs with the ewes through the summer, at least until the year that "Elmer" arrived. The summer after his birth Elmer was the only intact ram lamb to remain with the ewes and lambs. We kept watch over him as the summer progressed. Once he started to behave as expected of a ram, he was to be separated from the ewes. Unfortunately the little guy was smart enough to hide his talents from us but not from the ewes. Only by the end of August (at the age of six months) did we suspect that we had royally been had! So we separated him from the ewes. By then, however, we were nervous enough to push forward the scheduled breeding date for the adult rams into mid-September (rather than early October). When we did place the adult rams with the ewes it was quickly obvious that Elmer possessed well used talent. The adult rams were exceedingly bored, marking very few of the ewes. Later, after we had pulled all the rams in late-October, little Elmer managed to destroy a temporary wooden fence of solid construction and got back in with the ewes. He found a ewe lamb that he previously missed and thereby extended lambing by another two weeks.

The next "spring's" lambing ran from early December through the end of April. It will be forever remembered as "the lambing that would never end" or "the lambing from Hell"! Elmer had managed to successfully breed half the adult flock before he was four months old and all of the ewe lambs thereafter (a total of over 35 ewes). Because he did not wear a harness until mid-September we had absolutely no idea of when most of the ewes were due and which of them might or might not be pregnant. LESSON: We have ever since that year separated all of our intact ram lambs from the ewes at three months of age no matter what the books or experts tell us. We also invested in strong steel pens for our rams. Those pens have become bent by the butting of the rams over time, but we just turn the panels around for the lads to straighten out. How a few years later a ram lamb managed to weasel his way out of one of these steel fortresses is another story. As a footnote, many of Elmer's offspring from that year and subsequent years are still with us and are some of our best ewes. Elmer met an untimely death at less than three years of age due to a very unusual type of urinary calculi.

Conclusions

Every year we seem to learn more from our flock. We never seem to stop learning from them. They are our partners in a growing enterprise; we and the flock are constantly evolving. Our knowledge about each other grows and as a result, our facilities, pasture and genetics are slowly improving year after year. Hopefully, we have learned by our mistakes and miscalculations. I hope that we do not become complacent enough to forget that we will always have much more to learn. Hopefully this introduction strikes a few familiar cords with all shepherds. If it raises more questions, perhaps they will be addressed in the future. Thanks for listening!

Remembering Cimarron

January is the quiet time of our year. January of 1998 is little different than other Januarys in that regard. The ewes are half way through their pregnancies. Pastures are inaccessible due to deep drifts of snow. The flock seems quiet and comfortable in or near the barn with a good supply of hay. For the sheep, it is a time for eating, sleeping, ruminating and gestation. For us, it is a time for planning and resting up for the future. Shearing is less than a month off and lambing starts in five weeks. It is also a time for reflection. This year it is different than in previous years. The difference is due to the absence of Cimarron.

Cimarron was our first colored ram. Our first lambing in 1991 was exclusively the work of our old white ram, Monty, who quietly and efficiently bred every ewe presented to him. His lambs were all healthy, robust and white (as expected). For the next year we were in need of a colored ram to complete the initial phase of our plan to raise both colored and white Corriedales. Good quality, purebred colored Corriedale rams were few to be found in Wisconsin, except ones who were already related to all our colored ewes. It was by chance that, in the spring of our first lambing, we heard from a breeder of registered white Corriedales who (much to his surprise and chagrin) had produced a black ram lamb from the mating of two of his white sheep. In order to register a white Corriedale with the national breeders association, both parents of the lamb must, among other things, be purebred Corriedales, which means they must be white. The presence of colored genes is frowned upon and therefore not desirable. "Were we interested in the black lamb?" he asked. If the seller did not quickly find a buyer the odds were that the lamb would quickly "disappear" so as not to sully the breeder's reputation. Sight unseen, we said that we would take him (for a price not much higher than he would have fetched at the slaughter auction). It was admittedly a risky business buying a ram not yet two months old without even seeing him, but for us it seemed a potentially wonderful opportunity. If the ram lamb proved to be not what we were after we could still market him as a slaughter lamb and not loose much money on the deal. We knew that he was not related to any of our ewes or to any other colored ewes in the state. He was therefore the source of a new colored gene pool. All we could hope was that he would produce the type of lambs and fleeces that we wanted.

When we picked him up in early summer, it was obvious that the lamb was a bit wild. All of our lambs born in 1991 were given names beginning with the letter "C". For him we somehow felt that the name "Cimarron" had an appropriate ring to it. It was apparent that he was not used to much human contact, but he was not aggressive toward us. He spent the summer with a white ram lamb, Cuthbert, who was (like his father, Monty) the perfect gentleman. As the two grew, we eventually placed them with Monty and much to our relief and surprise, Monty treated them well. Cimarron at least became tolerant of us, but not often friendly and usually very untrusting. He was, above all else, a beautiful ram with a black, fine fleece.

That fall he was given the opportunity to breed our eleven yearling colored ewes. Given that none of them had been bred their first year and that Cimarron was an inexperienced ram lamb, he did well. After two heat cycles, all of those ewes joined Monty for clean up and it was obvious that Cimarron had missed breeding just a couple of them. Nonetheless we were pleased. The following spring we were even happier, as the colored lambs he had sired were wonderful. Cimarron was to become the foundation of our colored flock.

Over the ensuing years Cimarron contributed mightily to our growing flock. We retained a number of his ram lambs for use in our breeding program and for sale to others. Each of these rams have proven to be a wonderful, positive addition. Cimarron's lambs produced fleeces that generally exceeded their mothers' in quality and fineness. The fleeces were consistent, had a wonderful hand, and appeared in various shades from black to light gray. After a couple of years we began to see the appearance of a gene for large spots. Looking at bloodlines, it could only be attributed to Cimarron.

In the fall of Cimarron's second year we had just placed the rams with their various ewes when old Monty became seriously ill and subsequently died. The original plan had been for Cimarron to breed the colored ewes and Monty the white ewes. As we had no other rams at that point, Cimarron was, by default, entrusted with the entire ewe flock, by now numbering close to forty. To our delight, he did well with the large group, successfully breeding nearly the entire group. In the process, we jumped ahead with our breeding plans, as we had envisioned eventually crossing some or all of our white ewes with colored rams, in order to produce white ewes carrying colored genes. We had just not planned on it so soon!

14

Overtime Cimarron became friendlier. He especially enjoyed a good rub on his neck or behind his ears. At times like these he seemed peaceful and trusting, only to dispel the feeling a few minutes later when he would bolt from a seemingly innocent movement on our part. He was, above all else, a ram to never take for granted, but he never attempted to challenge either of us. As he got older he had medical problems here and there. In his fifth summer he managed to get his right ear caught on something (we still know not what). The ear was half torn off, a gushing bloody mess. One of our trusty vets did a superb job of stitching what was left back together. Once healed, the ear had a particular bend to it. Cimarron looked as though he had been in a barroom brawl; it somehow fit his character. Later that summer he had a particularly bad case of a respiratory illness, the exact cause of which the vets were never able to determine. It took a long while to overcome the illness and all the while Cimarron seemed to understand that we were doing our best to treat him. He would allow us to catch him and administer some painful medications without much resistance. But as soon as we ascribed this passiveness to his weaken state, he would proceed to clobber one of his pen mates. With his age, we began to consider a retirement for him. The pool of ewes for him to breed was becoming limited by his overall genetic contribution to the flock. In a perfect world he would have been allowed to live out his days with us. It was the least we could do to acknowledge his contributions to us. A perfect world did not, however, account for the darker side of Cimarron's personality.

There was indeed an almost sinister aspect to Cimarron. He was the dominant ram, but it was more than usual pecking order dominance. Even when he was ill no other ram would ever challenge him. He was not the largest ram, but he was the boss and was very willing to quickly remind any other ram of that fact. He was extremely powerful; he had the ability to destroy a wooden pen designed to separate him from the ewes and from younger rams. After rebuilding the pen using rough cut 2 x 6 and 4 x 6 timbers, we found that it only seemed more of a challenge for him to systematically destroy our handy work. We invested in steel pens which did contain him. It was, however, necessary to regularly reverse the panels so that he could beat back the bends to which he had subjected the steel rods. The metal fencing made quite a clatter as he repeatedly pounded upon them. It became his habit to start this din at feeding times in order to get fed first. It was a tactic that worked admirably for him!

In our second year, for lack of a good local shearer, I had learned to shear. I faired pretty well my first year (considering my advancing age and brief training). Cimarron was intentionally left as my last sheep to shear for the season. By the time I got to him I was relatively confident of my newly developing skills. Cimarron quickly took care of any over confidence on my part. He fought and kicked ceaselessly, managing to implant a bruise on my non-shearing hand that matched the base of his hoof perfectly and which took a month to heal. By the next shearing season we had located a wonderful professional shearer. I felt somehow redeemed when Dave had nearly as much trouble shearing Cimarron as I had experienced the previous year.

The afore mentioned characteristics can be ascribed to many rams in varying degrees. Cimarron however carried his darker side further than most. He seemed to take his role as boss to a fatal extreme. When placed with his group of ewes for breeding, he would become obsessively protective of them, constantly herding them into a tight group, even though no other ram was even near in an adjoining pasture to threaten his mastership of his harem. Any ewe that strayed away from the group was subjected to a beating. It

16

was a behavior that was fueled by the onset of the breeding season. After spending a couple of weeks with the ewes he would be more permissive of their wanderings, unless they were in heat. If he felt the ewe was ready for breeding, Cimarron would demand their total attention. Lacking that attention, the ewe often found she was subjected to a powerful butting. One morning in his first year as an adult ram we found a dead ewe in his group. The previous day Cimarron had been pestering her. After a necropsy, the vet could not find any other possible cause of death except probably a broken neck. We could only guess how it occurred. Over the years no other fatalities occurred, but we would occasionally find a ewe who had been recently bred who was also severely favoring a leg or two. This problem did not occur with ewes who were in with any of the other rams. On a couple of occasions we removed the gimpy ewe from Cimarron's breeding group to give here a chance to recover. It was a behavior we tried to ignore as best we could, but for which we were now constantly on the lookout. Had Cimarron not had the beautiful fleece character that he passed on to his offspring we probably would have been less forgiving.

In the fall of 1997 history repeated itself. When placed with the ewes, Cimarron appeared more frantic than ever. He circled them into a tight group and would challenge anyone trying to leave the group. The next morning when bringing the group their grain, we watched Cimarron bowl over a smaller ewe as she tried to wander a short distance to graze. By that time, we were more nervous than usual. At noon we noticed a ewe down in the pasture. When we got to her she was dead; her neck had been broken. It was Hilary, one of our favorites, and one of our few spotted ewes. It was a loss that we could not afford to repeat. We caught Cimarron and struggled to pull/push him back the quarter mile to the barn. The vet was called and we had Cimarron put down that afternoon. His group of ewes was traumatized enough by the event that it took them a week to settle in with another ram. It took the two of us much longer to overcome both the anger and the guilt we felt for allowing the loss to occur.

Now we await the first lambing in seven years to which Cimarron has not contributed. Our rage of last fall has subsided. Instead, we wonder where our future lambings will head. It has been the quietest and most peaceful winter in the barn in years. The remaining five rams still do not seem to be able to figure out who is in charge, but their disruptions are brief and directed toward each other rather than iron rails of the ram pen. Cimarron's genetics will be with us for years. One of his sons, Gabe, and a grandson, Iago, are with us and will hopefully add the best of his genes to the flock without the anger that Cimarron would so often express. Gabe has the rich black shade of Cimarron; Iago has the large random spotting that occurs infrequently in the Cimarron offspring.

It is also a season for starting anew, as we have a new colored ram lamb, Ironsides. The fact that he was a colored lamb was unexpected. His mother is Ducky, a white ewe, the daughter of old Monty and one of the original colored ewes, Beatrice. Ironically the year in which Ducky was conceived, Beatrice was one of the colored ewes that Cimarron, as a lamb, failed to breed. What made Ironsides' color so surprising was that he was sired by Greenup, one of our registered white rams. It was a sign that although he was a registered white Corriedale he carried colored genetics. As we have closed our flock, this was one of the few chances that we could have to introduce new colored genetics without using artificial insemination. We have the beginning of a new colored bloodline, one which does not contain any Cimarron genes.

So it is quiet in the barn. The joy of shearing is close, that wonderful time almost like opening Christmas packages. The excitement of seeing new lambs, especially those that Ironsides will produce, is still not upon us. The threat of physical harm seems much more distant than it has in many years. Nonetheless, it is in these quiet reflective moments when I also miss Cimarron the most.

Cooler Near the Lake

So...have you ever tried to explain global weather patterns to a flock of sheep? I tried this morning and failed miserably. On further reflection I realized that correctly explaining the weather to anyone, let alone a sheep, is an impossible task. Because of that oversight I believe I also learned a few ovine curses that I previously had not heard. I have now been shepherding long enough to believe that sheep have a language with which they seek to communicate with us. I have yet to learn the key to that language to be able to fully respond to the flock, but I am trying! Here is what I had hoped to communicate to the girls and their lambs this morning.

Wisconsin has not been spared the unusual weather that has blessed and/or cursed much of the globe this year. After a mild winter we entered a warm and early spring. Prospects were high for lush green pastures all ready for grazing at record early dates. In fact, we had the ewes and their lambs into a rotation on the pastures in late April, a full two to four weeks earlier than in the previous eight years. It was a joy to have the flock out so early. It is hard to tire of the sight of ewes and their young lambs enjoying the freshness of spring on new pasture. It made it easier to train the lambs to electric fences as their youthful curiosity was still very strong and their inquiring noses and mouths were quite moist. The overall health of the flock had benefited by the fresh, clean environment. A new, fresh pasture everyday was quite obviously tasty and nutritious.

As we progressed into May, it became obvious that one cannot generalize about weather for an area as small as Wisconsin. Portions of the state have had so much rain that the crucial spring planting was delayed. Other areas have been so dry that planting was completed at record early dates. In our little corner of the state, we began wet and rapidly progressed to very dry. By the end of May, drought had become a common word, especially for a period in which normal rainfall is abundant. We have already rotated through most of our grazing pastures twice and the pastures are beginning to act as if it were the usual late summer growth slump. Hay is being cut a month ahead of time, and much of the cutting is being done because the alfalfa and other legumes are starting to wilt. So we are scrambling to find enough hay for next winter and possibly to feed for part of the summer in the event that the pastures do not come back.

This weather/geography lesson is further complicated by very local conditions. We reside on the extreme eastern edge of a long (65 mile), narrow (1 to 10 mile) peninsula that juts into Lake Michigan, forming Green Bay to the west. The shore of Lake Michigan is a quarter of a mile to the east (you can view the water from the top of the barn roof if you are brave enough!). One's location on the peninsula can drastically influence weather and plant development. If you live near the bay of Green Bay, the warm westerly winds heat up the shallow bay water much more rapidly than the deeper waters of Lake Michigan on the east. If one thing is predictable, it will be a forecast calling for temperatures "cooler near the lake". With rainfall having been the same for the peninsula in general, it is amazing to see how much drier the hay and pastures are just to the west a distance of one or two miles. I have tried to explain this to the flock, but they do not appear to care. For them, the concept of "it could be a lot worse if they lived a mile farther west" has little relevancy. When looked at from their perspective, the concept of "a mile farther west" is foreign; most of them have never in their lives been more than a quarter mile west of here.

Our pastures have held up pretty well, considering the lack of usual spring rains. Primarily this is due to the cooling winds off Lake Michigan. Nonetheless, we are starting to alter our grazing rotation to compensate for the lack of rain and the lack of normal pasture regrowth. Yesterday the ewes and lambs were given a larger than normal paddock to graze, in hopes of it not being overgrazed by day's end. Apparently this offering was to the liking of the entire flock. In the morning they greeted the pasture with gusto and the silence of contented munching. You know that it tastes good when the sheep start grazing without any editorial comments, save for an occasional call to or from a lamb.

This morning dawned with an unusual occurrence: rain and thunder. Only the second rain in the month, it amounted to only two tenths of an inch when finished, but contained more than its share of electrical activity. Not being too keen about handling an electric fence in the middle of a thunderstorm, we opted to leave the flock in the barn for the morning or at least until the lightning had passed. There was still baled hay they could be offered to munch on, even if it generated a general amount of disgusting comments from the ewes. By the time the weather had cleared up we checked yesterday's pasture and found that there was still sufficient grazing to cover a half a day's graze. So by midday the girls, their lambs and their shepherd headed out to pasture with eagerness and anticipation of good things to eat. No sooner had we reached the designated pasture than the entire flock let out a terrible clamor. They stood there, all 155 of them and complained bitterly. They did not rush to the far end of the pasture as they usually do when first arriving. ("After all we were there yesterday.") The din became louder by the minute. They did not start to look for the good grazing. ("What the heck, we ate it yesterday.") All they did was complain. ("Why the heck did you give us left over pasture?") The complaints rained down upon me seemingly forever. "Complain" is too mild a term, as I was being cursed in "ovinese". Were we not separated by an electric fence, I began to feel that they were contemplating a mob action against me. Even after I had left them and walked the quarter mile back to the barn, I could still hear them.

I had not suggested it to them in the state they were in, but I felt strongly that they were acting a bit spoiled. Later I tried to convey that notion to them. After all, they were a lucky flock; they could be grazing a mile or two to the west where the pastures were drying up and it was hotter. Somehow I seem to have failed with my description of what was going on weather wise and why they might have a little less desirable grazing if this continued. Even the concept of "Cooler near the Lake" did nothing to impress them.

Should this dryness continue, it will mean an early weaning for many of the lambs, along with their earlier than normal marketing as feeders. The difficult decisions will have to be made as to which of them gets to stay as replacement ewes and who will have to go. Will we keep fewer than planned due to lack of feed? Who will we choose and how many? All these thoughts are just for the sake of conserving enough grazing to get us through the abnormally dry times.

This year was to be the one in which we expanded our ewe flock, perhaps by 20 or more. The lamb crop is a nice one, with a good number of potentially wonderful fleeces appearing. We can rationalize the decision to not keep a number of lambs if they do not meet our standards or if their numbers could not be normally supported by the acreage. However, when the usually lush growth of pasture rapidly deteriorates and the prospect is for six or seven more months of grazing with nothing to graze, it clouds the selection criteria dramatically. So for the time being, we will try to be thankful that at least it is cooler near the lake and then hope that the rains will soon materialize.

The Long and Short View

Being relatively new to writing for a print publication, I still find it difficult to realize that what I write now in early September will not be read by anyone until mid-December. We have certainly become spoiled by instantaneous electronic communication, but we have perhaps lost a bit of measured thought by not taking the extra time to write, think about what was written and then re-write a while later. The act of writing something for the future is similar to the act of planning for the upcoming season for the flock. As with any agricultural enterprise, there is always a need to take a long pause every once in a while to look at where the farm is headed into the future.

In this vein, by the time this piece gets to press, our ewes will have been bred, and sheep and shepherds alike will be into our winter mode. Yet as I write, we have barely made our final decisions as to which lambs to keep as replacements, along with which ewes to breed to which rams.

Mother Nature has continued to play a dramatic role in our decision making process. The summer continued the erratic patterns with which it started. We have been cursed by drought, saved by heavy rains at the end of June, only to return to drought for the remainder of the summer. As we feared, the on-again, off-again pasture growth throughout the summer has had an influence on our replacement ewe selection. Faced with rapidly disappearing grazing opportunities, we sold many more of our lambs earlier than normal as feeders. Doing so did not give us the time we would like to properly evaluate the ewe lambs as potential replacements. I am sure that we may have sold a few lambs who might have developed into something very special had they the benefit of a couple more months growth on the farm.

We are still gambling that the weather will treat us more kindly next year, as we are keeping nineteen more sheep than we over-wintered last year. We have managed to scrounge up enough hay to be able to feed about ninety sheep for the winter (if it is a normal winter!). Lacking any significant rain in the next two to three weeks will necessitate a purchase of additional hay to supplement the normally abundant pasture of early fall.

We have just made the final selections regarding which ewe lambs and ram lambs we will be keeping. For a couple of the lambs, the decision was made not more than a week before we made our final trip to the sale barn. We have noticed an interesting behavioral pattern amongst the recently selected lambs. All of our breeding stock is jacketed year round. The jacketing serves to keep the wool exceedingly clean. In the case of the colored sheep it also significantly limits any sun-bleaching of the wool. As soon as we have made a decision to keep a lamb for the future it receives a jacket. Thus some of the lambs are jacketed at quite an early age (perhaps as soon as eight weeks of age). In the case of some of the others, the decision to keep is made much later. As can be guessed, this pattern of jacketing also requires that we have quite a large wardrobe of different sized jackets so that we can accommodate the very small to the full-grown sheep, plus increasing the sizes as the lambs grow and the adults add more diameter as their fleeces grow. Regardless of when the decision is made to keep the lamb, it is then

26

outfitted in a coat. The coat style has not changed for many years, but the lamb behavior upon receiving the coat has evolved. At the risk of being anthropomorphic, for the lambs the act of getting a jacket seems to have become a rite of passage within the flock. We sense that the lamb "knows" that it is staying with the flock when it receives a coat. Wearing a jacket for the first time involves new sensations for any sheep, as there are straps fitting loosely around the base of their back legs and a gusset around their neck. Now the lambs are much less frantic when fitted with their first jacket than they were a few years ago. In some cases the lambs behave as if they have worn a jacket all along. They also tend to become much friendlier toward us once they have received the jacket.

The process of selecting replacement ewe lambs is always frustrating; it seemed especially so this year. We hope to maintain a diverse selection of fleece colors, without knowing which shades of fleeces will sell most rapidly in any given future year. In addition, the colors/shades of our lamb fleeces frequently change markedly after they have undergone their first shearing. For example, will the really black ewe stay black, or will she turn to gray in a year? Will the spotted lamb's spots remain distinct or will they merge into a subtle marbling? It would be wonderful to know more about the complex colored genetic make-up of each sheep, not just in terms of its coloring patterns, but its tendency to retain or loose its original shade.

Mating the ewes to a specific ram adds to the genetic puzzle. We will be using "Ironsides", one of last year's ram lambs, as a primary breeding ram for the colored ewes this fall. His shading, marking and colors are dramatically different than any of our previous colored rams. We bred him last fall to some of the ewe lambs, so that we have a taste of what his genetics will contribute, but it was a very small taste. Thus for at least one more breeding season, his contributions will largely remain a mystery. It is the mystery however that makes the arrival of the lambs especially exciting each spring, perhaps more so this coming year than normal.

Despite hours of reading and re-reading wonderfully detailed articles on colored sheep genetics in Colored Sheep and Wool[1] and Black Sheep Newsletter[2] we still dream of the year that we will be fortunate enough to attend the World Congress on Colored Sheep to share ideas and gain further insights into colored sheep genetics. We may try to photographically catalog what each member of the flock looks like immediately after shearing. We often regret not having such a record and instead having to rely on our memories and written notes. A good "barn guide" to colored sheep genetics would certainly be a rewarding manual, especially if it could contain standardized photos of the various patterns as they appear in each breed. Add to that, a discussion of which patterns are dominant over another; spice it up with some research into the genetics of fading versus color retention, and it would become a fascinating piece of literature. Anyone out there want to get something started along these lines?!

For us it is back to planning for the upcoming fall breeding based upon our past experience and results. By the time this letter is "delivered", we will hopefully have made all the right decisions and Mother Nature will have blessed us with sufficient moisture.

---

[1]Erskine, Kent, ed. 1989 Colored Sheep and Wool, Exploring Their Beauty and Function, Ashland, OR: Black Sheep Press
[2]Black Sheep Newsletter, Scappoose, OR

Uninvited Guests and
Other Distractions

It is amazing how seemingly minor events can disrupt a breeding season with the sheep. Last year went so smoothly and effortlessly that we were spoiled for any subsequent year. In 1997 our three adult rams and two ram lambs each flawlessly performed their assigned tasks in less than a month. The ewes were equally receptive. By the end of that time, each ram appeared very bored, spending increasing amounts of time longingly looking over the fence toward the next nearest breeding group, wondering whether life was more exciting with that group. There is no better sign than this that all of "his girls" are pregnant, at least for the moment. The payoff came this spring with a lambing that lasted three weeks. It was an intense time for ewes and shepherds, but well worth it. If we were going to spend lots of time in the barn, we might as well be busy during that time! Having a group of ninety one lambs, all of virtually the same age, continued to pay labor and feed dividends throughout the spring and summer.

The breeding season this fall was a marked contrast to the previous year's. Sheep have an uncanny way of quickly negating any overconfidence on the part of their shepherds. This fall a series of seemingly unconnected events ended up disrupting the hoped for smooth performance of ewes and rams.

We started things off poorly by creating a lot of noise and distractions for the flock. An old concrete silo came with our farm when we purchased it. The silo was attached to the south end of the barn. The silo was in poor, unusable condition even in 1983. Over the years it had remained a memorial to the cows and dairy farmers who preceded us here. It would be of no use to us, since we never planned to produce enough silage to justify putting it back into operation. At times we fanaticized about building an observation platform on the top of the silo. It would have afforded a magnificent view of the entire farm, plus Lake Michigan to the east. It would probably have offered the most spectacular night time view of the stars on a warm summer night. On the other hand, even without a "bridge", the silo was not a totally inanimate object. Its nearly empty interior resonated with the pleasant cooing of generations of pigeons who called it home. Moisture and Wisconsin winter temperatures were slowly bringing down chunks of "rotten" concrete. Initially these chucks were small, but over the years they became larger and fault cracks were becoming more evident.

Pigeons or not, it was time to take the silo down before a large piece of rubble did severe damage to a shepherd, to a few sheep or to the barn. Since the removal of the silo would leave a large void at one end of the barn, we decided that it would be an opportune time to add on to the barn. By the time all the work was scheduled, it was fall. As with many such projects the scope of the job grew, seemingly by the day. It became apparent that the only "safe" way to bring down the structure was with a few strategically and professionally placed sticks of dynamite. Even with the flock in a distant pasture, the rumble of the dynamite blast was disruptive. Over the course of a month the removal of the rubble and subsequent construction of the barn addition caused a visually stressful and noisy environment for the flock. It became obvious that placing the rams with the ewes amidst all the chaos would lead to disruption of their normal breeding performance. The flock was not as comfortable as they should have been and it showed. Breeding time was upon us. We could not delay any further if we were to have any hope of getting the new lambs out on pasture as soon as the grass started to grow next spring.

Discomfort seemed to feed upon itself. During the silo destruction and immediately prior to placing the rams with the ewes, the adult rams were behaving especially macho amongst themselves. The more noise and clatter from the construction, the more the rams seemed to fight with each other. There were a couple of days of serious head butting that ensued. A yearling ram was claiming a higher status within the group. The unfortunate recipient of this status adjustment was his father, Greenup. We separated him from the others to avoid further injury. Once we had placed Greenup with his ewes it became apparent that his separation had come too late. He was not able to sustain a mount on a ewe without his hind legs giving out. It appeared that he had suffered either a pinched nerve in his back or strained muscles in his back legs. It was obvious after a few days with the ewes that he needed a rest; he was not vigorously and successfully breeding his ewes. Using a marking harness on the rams had further confirmed it. At best his efforts were few and doubtfully successful. So off Greenup went to the barn for rest and rehabilitation, while his ewes marched off to another of the rams, Gilbert. After a few days with Gilbert it was apparent that the ram effect had caused nearly all of Greenup's ewes to cycle in the first few days they had been with him. Lambing would now be extended at least another 17 days or more.

The next disruption a few days later seemingly came from out of nowhere. In Gabe's breeding group, three pastures and an eighth of a mile away, we were greeted one morning by the strong odor of an uninvited guest. There was no skunk to be seen, but the odor seemed to accompany us as we moved the sheep to a different part of the pasture. At least one of the ewes had decided to make the acquaintance of a skunk the night before. Over the next few days we managed to separately sniff each sheep in the group. It must have been a bizarre scene in the middle of a large pasture: two middle aged humans wandering amongst a group of sheep, periodically stooping to smell each one. Miraculously only two of the ewes had been directly sprayed, while the remaining 17, plus Gabe had been spared. Of the two ewes, one had already been bred (another plus for the marking harness), while the second had not yet been marked. Changing their jackets helped somewhat to improve their smell, but it did not eliminate it. The second ewe would remain unmarked throughout breeding. Either she was not cycling, or the skunk smell masked it from the ram, or perhaps Gabe just could not tolerate her choice of perfume.

In the past we have had other similar potentially disruptive encounters between the sheep and other animals. The two previous years found us pulling porcupine quills from the muzzles of a number of ewes. One fall a pair of Canada geese took up residence in the middle of one of the pastures. Unlike their experience of getting to meet the porcupine, the sheep gave the geese a very wide birth. It was comical to see two geese peacefully grazing in the middle of a pasture, while a flock of sheep huddled in a far corner, seemingly afraid to graze anywhere near them. That fear was overcome or at least forgotten this fall. A number of migrating geese spent quite a few days landing near the sheep and then grazing amongst them without any fear being shown by either species.

The final disruption directly involved Gabe, who, after three weeks with his ewes developed a severe limp with one of his front legs. An examination found an injury to the top of the hoof which had become infected. Our vet got the hoof cleaned out and bandaged up. It was decided that Gabe needed a rest to recuperate, so it was off to the barn to be replaced by a rejuvenated Greenup. It should be noted that despite obvious discomfort, Gabe was still hard at work that last day, managing to successfully mount a couple of ewes with just one good front leg.

Traditionally, we bring all the breeding groups in from pasture before the beginning of the last full week of November. Wisconsin's gun hunting season for White-tailed Deer begins then. The noise of early morning gun shots near and far is its own disruptive force and it never fails to raise the anxiety level among the sheep. So by the Saturday prior to Thanksgiving everyone was comfortably housed in their newly expanded barn, waiting for the shooting to subside so that they could once again get back outside. Normally this date is when we close out the breeding cycle by putting the rams back in bachelor quarters. By this time we were quite nervous about how successful our breeding program had been. So after two weeks of rehabilitation, Gabe was ready to go back to work when the flock returned to the barn. He got an extra three weeks with all the adult ewes to hopefully clean up. It has proven a wise decision as there have been a few ewes from each group that he has remarked and, hopefully, successfully bred in the quieter and more peaceful environment.

Lambing will indeed take longer than expected next spring. We are just thankful that peace has returned to the farm. We are equally as thankful that we have an old reliable ram who, given the chance, we can count on for assuring a productive, fruitful lambing to come.

Building Fences and
Watching Grass Grow

It is March and Mother Nature has yet to release the soil on the farm from her frozen winter grip. Yet, for once, we actually hope to plan our fencing needs ahead of the grazing season and prior to the season for building fences. Last year we made a big decision (at least for us) to significantly expand the size of our flock. If the sheep were to pay their way year in and year out, they needed to generate a greater inflow of cash.

The flock's ability to grow in size was dependent on our meeting three needs at the farm. First, we needed adequate housing facilities for the worst of winter and to assure a safe lambing season. Second, we needed increased pasture acreage to allow the sheep to graze in a sustaining fashion from beginning to end of our grazing season (usually late April to late November here in northeast Wisconsin). And third, we needed to become more self-sufficient in the production of our stored winter feed, rather than purchasing a substantial percentage of our baled hay.

We have seemingly met our housing needs. The removal of the old silo at the end of the barn and the resulting hole in the ground proved to be a good site for an add-on to the barn. The majority of the construction was completed before the dead of winter. A portion of the flock moved into their expanded winter quarters prior to the first storm of the winter. Now we await the imminent arrival of this year's lambs to test the facility to its fullest.

Our second and third needs (pasture and a totally home grown supply of hay) will hopefully come with the permanent enclosure of our last 30 acres of tillable ground. When we purchased our farm of 80 acres, about 40 acres were under cultivation. We decided to continue to rent out the tillable ground to a friend and local dairy farmer, knowing that we had the assurance of rental income plus the careful stewardship of our land. Of the remaining acreage, about 15 acres had been cleared of trees but had never felt the blade of a plow. It is rough, uneven ground, often dominated by large rocks and boulders deposited during the last Ice Age. It was this cleared, rough ground which became our first grazing area. It was land wonderfully suited to the agility of grazing sheep, but a terror for an attempted tractor crossing. As our flock grew over the first few years, we gradually enclosed this rough ground with permanent perimeter fence, eventually completing four separate paddocks.

As the flock grew, so did their grazing needs and we therefore expanded onto some of the tilled ground. We fenced a five acre plot that was already planted to hay. An additional five acres had been devoted to a cherry orchard and it was fenced, in order to exclude the local deer population and to permit occasional grazing by the sheep. Presently, the remaining 30 acres of tillable ground is planted in hay. We have been grazing at least five acres of that land, relying on portable electric netting for fencing. The remaining 25 acres we have continued to rent to our dairy farmer friend. The logistics of this portable fencing has become frustrating, as our resident, ever-growing deer herd manages to frequently entangle themselves in it over night. Hence, we are about to embark on fencing the final 30 acres with a permanent perimeter fence and more user (i.e. shepherd) friendly movable interior fencing. This change will allow us to mechanically harvest some of the hay on a rotation for winter feed and allow the sheep to harvest the rest for spring, summer and fall forage. An additional benefit to be gained will be parasite control. By cutting a section of the pasture for hay for a year, the cycle of internal parasite reproduction on that section should be broken and should permit worm free grazing on that section the next year.

Our system of fencing has evolved over time and has been inspired by many sources. It is a system that works well for us and may not be at all suitable for the next shepherd. We learned by our own early mistakes and will probably continue to do so with mistakes that we have yet to make. Each paddock's perimeter is enclosed by permanent, electrified high tensile fencing, usually consisting of six wires. The bottom four wires are spaced at six inch intervals and the upper 2 wires at larger intervals for a total height of 42 inches. Above all else, a quality, high powered, low impedance fence energizer is essential. It must have the capacity to consistently deliver a strong charge to the most distant pasture under all sorts of conditions.

The pastures are laid out either according to natural divisions in the landscape or are parallel to the original stone fences. The stone "walls" were (and still are) the depository for the annual spring harvest of stones, which "grow" with great profusion in this part of the world. Within the boundaries of the perimeter fences, the sheep graze smaller sections of the paddock, usually on a daily basis. The size of the subsection is determined by the amount of forage available to eat, its rate of growth and re-growth, and the number of mouths available to do the harvesting. It becomes an interesting chess match for the shepherd/grazier (me) to apportion the proper daily acreage while trying to anticipate Mother

Nature's next moves with rain, sun and temperature. It is a challenge both exciting and often very frustrating.

The subdivision of these paddocks is achieved using removable, tread-in posts that carry two strands of electrified polywire that can be quickly rolled up and then unrolled in the next configuration. However, a bucolic pastoral scene is not achieved without some ovine education. We have learned that our sheep only respect these two small wire subdivisions if they have had a proper introduction. Inappropriately educated sheep are a real pain to deal with when it comes to moveable electric fence. We start education with the lambs on the first day they are introduced to pasture. Instead of the usual 2 wire fence we will set up three wires, more closely spaced. Often one or more of these strands are the more visible polytapes. The sight of a half inch wide white strip fluttering in the breeze is intriguing to the curious young lamb, at least until it touches a moist nose or mouth to the electrified tape. The lessons are rapidly learned; by the end of two or three days nearly all the lambs give the fence a wide birth. It also seems to help to have well trained ewes who never challenge the fence. The lambs of such veterans do not venture too close to the fence without their mother. On occasion, we will have a lamb who will learn that by lowering its head and ears and quickly scooting under the bottom strand that its wool will insulate it sufficiently from a shock. If this behavior is not quickly cured, it easily becomes a habit. Such a "bunch quitter" will tempt others to join in. If we are not able to rapidly change the lamb's beliefs, the behavior becomes prime grounds for culling. Luckily we have only had a couple such lambs over the course of recent years. Their rare appearance does not warrant a change in our fencing practices.

During the grazing season, my chore time takes perhaps an hour each day to move the fences and sheep to a new location. It is an hour well spent watching the sheep and lambs, determining whether they all appear healthy and satisfied. It permits me to assess the area grazed on the previous day. (Was it too small or too large? Will it recover more or less rapidly? Is the desirable mix of grasses and clovers still present?) In addition I will probably walk through the adjacent pastures and assess their status as far as grazing capacity for the future. Lastly, there are some important intangibles associated with these chores. It is a very quiet early morning hour when I can reflect on the day and the satisfaction of shepherding such a trusting group of animals and friends. To be able to stand back, to watch the sheep peacefully graze and to listen to them busily munching down mouthfuls of fresh grass and clover is truly one of the deeply satisfying moments of shepherding.

So, this year as soon as the frost is out of the ground and the grass is growing, the sheep will again be out where they belong grazing on the pastures while this shepherd will be out setting corner posts, stringing wire and dreaming about what the flock will look like on the new ground.

Reflections on Motherhood

It is mid June 1999 and the rains continue to fall on northeastern Wisconsin. What a contrast to a year ago! The pastures are lush and green. In mid-June they are still growing so rapidly that the grazing of the sheep just cannot keep up with it. In contrast to a year ago, we cannot make hay without it getting rained upon. It is a time to wait for a dry spell of sufficient length to cut, dry and bale hay for winter. In the meantime I can watch the sheep and lambs getting wet one more time as they graze. It gives me an opportunity to see how well each ewe and her lambs are doing and to observe parental relationships as they develop.

We lost four especially good mothers this spring due to a variety of mostly unrelated circumstances. They will all be missed, not just from a production standpoint, but because they were all, in their own way, our good friends.

Girlie was the first to go. Early in her pregnancy it became apparent that she had developed a flank hernia. It is a condition not realistically repairable in sheep. As her pregnancy progressed she had increasing difficulty with mobility. We nursed her along as best we could. We eventually gave her private quarters and individually served feed. When it came time to lamb she did well, considering her condition. Both of her twin ram lambs were delivered without too much difficulty and both were in good health. Despite the private care, Girlie had called up too much of her bodily reserves to support either the lambs or herself. Shortly after lambing she succumbed to milk fever. The combination of conditions was too much for her to overcome.

Girlie was always the model mother, extremely devoted to her lambs and their care. She never failed to develop a strong bond with any of her offspring. Those of her ewe lambs that we kept always slept close to her, even as adults. Both of her twin ewe lambs from the previous year, Jerusha and Jeunefille, are still with the flock. They were both successfully bred and each lambed prior to Girlie this spring. When Jerusha and Jeunefille each went into labor, they settled down very near their mother. Each of them carried on a quiet, yet serious conversation with Girlie as delivery progressed. While she may have shared her lambing secrets with her offspring, Girlie managed to hide them from us regularly. In more than one year we were highly concerned that her lambs were not getting sufficient milk, as we never saw them nurse and they complained often enough for us to check Girlie's milk supply which did seem to be lacking. After a couple of days of this interaction, the scales told us that we had been had. Girlie consistently waited until our backs were turned before letting the lambs nurse! Girlie's relationship with us was much the same as with her lambs. She was trusting, devoted and always a little quiet and shy. A lot more than her beautiful fleece will be missed.

When we lost Amanda, it was not expected. She was one of our oldest, original ewes; one of those we had purchased as a lamb. Unlike the trust that Girlie shared with us, Amanda never truly warmed to us in her nine years. Her philosophy seemed to be "stay as far away as possible without making it obvious". She did not survive the last hours of her ninth pregnancy for reasons that we will never know. She appeared to approach motherhood as a

40

business proposition: take good care of the lambs, see to it that they are alright, but no need to get too personal or intense about it. It seemed to work for her. In eight lambings she raised all 15 lambs without much fanfare or problems. In her quiet, deliberate way she was a major producer. She asked for very little, but her contributions to the flock were great.

Gabbie was the victim of meningeal worms, deposited on our pastures by the hordes of white-tailed deer that overrun our farm. Difficult to control, we thought that we had arrested the spinal injury that a worm caused late last summer. Only after she gave birth this spring did she show major signs of problems with her back legs on the long walks to and from the pastures. She was still young, having only experienced four lambings. She took on motherhood with gusto, authority and joy. It was an attitude that seemed to permeate everything she did. Her family units were seemingly loosely bound. Rarely did you see her with lambs close at hand. However, a couple of "Maa's" quickly assembled her troops. It is unfair to judge how close she and her lambs would remain, as she produced a long string of ram lambs; the opportunity to have a daughter, who could remain with her, occurred only once. She was an extrovert who relished company, human or ovine. Nevertheless, her lambs always came first for her. She seemed to pour everything she had physically and spiritually into raising lambs, be they triplets or a single.

Gidget shared the same meningeal worm problem as Gabbie's and also the same fate. In nearly all other aspects of life they shared little else in common. Four and a half months into her first pregnancy, Gidget was still bouncing around like a two month old lamb. We often wondered what that first lamb thought about the exciting ride! She did not complete her maternity homework assignments (too busy bouncing off walls?) When the first lamb arrived she did not know whether to love it or beat it to pieces. A week in a stanchion brought her to her senses; she in fact became inseparable from the lamb. Her next lambing began much the same way except with twins, one of whom she remembered to love, and the other one to abuse. A one day refresher course in the stanchion and a verbal threat ("You will not get another chance") was enough to get her on the right track again. As with the previously mentioned ewes, much of her mothering traits seemed to be inherited. Gidget's one and only daughter was just as much of an over-energized "air head" as her mother had been as a lamb, but she must have heard about the stanchions from her mother, as she started out willingly accepting her first lamb this year.

Mothering ability and style is highly variable within our flock. Nevertheless, the styles and talents of each ewe seem to pass on from generation to generation. It is unfortunate that the loss of four very distinctive ewes is the catalyst for reflections on their talents and personalities. Each of the four will be missed, each in their own special way. They were all major contributors to the profitability and improvement of the flock. Each had their own special fleece. Each produced a good crop of lambs year after year. Their loss will not just be financial. They were all good friends in their own personal unique way, and for that they will be especially missed.

The Question

"Why do the sheep wear jackets?" This question and a multitude of similar variations outnumber all other public inquiries we receive at our farm. Were we able to charge a fee for each time the question is asked we would be fiscally sound for life!! What is at the root of this issue? Why is it so often asked of us?

We raise white and naturally colored Corriedale sheep. Our adult flock usually numbers about 120 to 130. Our farm is located in Door County, Wisconsin, which happens to be a popular tourist destination, especially during the spring, summer and fall months. Our farm is also located next to one of the five State Parks in the county. "Our park", Whitefish Dunes State Park, boasts a grand sand beach and the tallest sand dunes on the western shore of Lake Michigan, all draw tourists, day trippers and locals alike. As a result, there is more daytime traffic on our country road than one would expect. Throughout the grazing season, which roughly corresponds to the tourist season, part or all of our flock is usually pastured within view of the passing road.

This area is traditionally bovine dairy country. The sight of a herd of large Holsteins is still much more common than the sight of a flock of sheep, let alone jacketed sheep. Few shepherds jacket their sheep due to the cost and the maintenance logistics that they entail. All of our adult sheep are jacketed year around and our replacement lambs graduate into clothing at two to three months of age. As a result, the sight of a jacketed ram or a jacketed ewe with her jacketed lambs by her side is the norm as you travel past our place. Admittedly, it is not the most common agrarian view to come across a flock of sheep here in cow country, especially sheep all decked out in coats.

Ever since we jacketed the flock, there has been a steady stream of traffic into our farm yard with "THE QUESTION" poised to be asked. "THE QUESTION" has many variations. Most often it is "Why do the sheep wear coats"? The coats are also described to us as jackets, bags, paper sacks, body suits, blankets, slickers.... If there is more than one person in the party asking "THE QUESTION", there is often an "expert" who is also "sure" of the answer. On rare occasions the "expert" will have a good explanation; most times their thinking mystifies us. Often the "expert" will state that it is because the sheep have just been sheared and they are cold (a suggestion usually offered on a warm summer's day).

When we finally offer our answer, it is always in two parts: 1) the jackets keep the wool clean and 2) the jackets keep the naturally colored wool from fading in the sun. If the questioner seems interested, we then describe why clean and non-faded wool is important to us and to the people to whom we sell it. The concept of increasing the value of our wool seems to be one understood by most people. Often the ensuing discussion can be detailed and fruitful. We may end up describing the process of jacketing each sheep. Folks are often surprised by our need to have a large supply of jackets that we can use as the lambs grow or as the adults increase their body diameter with the growth of up to four inches of wool. Perhaps in those situations, we have contributed in some small way to the greater knowledge of the questioner. What is perhaps more disturbing to us is to respond to the basic question and then have the questioner satisfied without any further detail. We have, as a society, become so distant from the sources of our food and fiber that the concepts of how it is raised and produced are now totally foreign to so many. It is refreshing to discuss sheep and wool with a person who genuinely wants to learn more.

Our experience with jacketed sheep has been educational for us as well. Not all sheep breeds and not all climates are equally suited to the use of jackets. We have found that our Corriedales' fleeces respond well to being covered here in our Midwestern climate. We have not experienced pilling or felting of their wool. The resulting clean fleece is a joy visually and to spin.

The coats generate additional work for us as shepherds. Coats seem to be a magnet for any small projection. My fence building techniques have been modified and "smoothed" for the sake of the coats. I have also pruned any small scrubby bushes or tree branches in our pastures on which the coats could catch. We constantly monitor the fences in and around the barn for any screws or nails that may be working loose.

Finally, there is coat cleaning and repairing (which is Gretchen's least favorite task!). It is such a large job that we have a washing machine devoted solely to the job of washing sheep coats. Because each of the adult sheep passes through at least three sizes of coats during a season and the replacements lambs often go through four or more sizes, there seems to be a continual mountain of sheep coats to be washed and repaired. We have found that one coat seems to last about three to four yearns. Some of the smaller sizes which are used for less lengths of time each year last longer. This is true also for the largest of sizes which only end up being used toward the end of each "fleece year". Conversely, some of the most popular moderate sizes, which may be in continual service with only a few days out for washing, may not last more than two or three years.

We have fantasies about creating a designer line of sheep wear, perhaps with logos or advertising slogans. When these outfits were complete all we would have to do is train the sheep to line up in the correct formation so that the slogan could be read by the passing tourists. On further thought, perhaps we should just be satisfied that in a small way we are contributing to better understanding of sheep and fiber production whenever we are asked "THE QUESTION".

P.S. Just as I finished writing this piece, we were asked a new variation: "Why are the sheep wearing sweaters?" Think about that for a moment in terms of the cast of characters in this drama. How would you answer???

Musings on Selection of Sheep

It is again December and quiet times have come to the farm. The ewes are all bred (hopefully) and are well into the middle of their pregnancies. We await the first snow of the season, already long overdue. There is still time to take a few extra breathes before shearing. The time gives me chance to reflect on some thoughts that recently appeared in <u>Black Sheep Newsletter</u>[3]. The issue contained a number of philosophies related to the selection of sheep for a flock. I believe that it was the intent of the article to begin a dialogue on the subject within the <u>Black Sheep Newsletter</u> community.

I apologize if I over simplify or misinterpret the thesis! It sets out three main points, all related to the development of the flock. They are: 1) there are no bargains in life, beware the bargain sheep, 2) thoroughly research a breed before you commit to it, and 3) develop a philosophy to employ when selecting your animals (either from within or without your flock). It is the final point of the three upon which I would like to elaborate.

How do you decide which lambs to keep and/or which sheep to bring into your flock? I propose three major areas of

---

[3] Black Sheep Newsletter, Fall 1999. Black Sheep Press, Scappoose, OR

concern. They relate to: 1) financial goals, 2) personal goals, and 3) genetic preservation. In my opinion no one of these three areas is more important than the next and all three are tightly interwoven with each other.

Financial Goals

What are the financial goals of the shepherd for his or her flock? These goals will have a bearing on decisions as to selection of individuals. The discussion focuses on two extremes regarding selection philosophies for the flock. At one end is an attempt to preserve a line that characterizes your flock with a certain "look". It emphasizes uniformity, for example, in fleece fineness, color of fleece, body size or build for members of the flock. At the other extreme is a flock emphasizing diversity within the breed. For financial goals either extreme may be appropriate. It all depends upon your market and how you wish to develop it. If you wish to make a profit (or at least cover your expenses!), you will need to select your sheep based upon what you are able to market. Are you selling breeding stock? Are your buyers looking for an animal that does justice to the breed standard, or which produces a particular type of fleece or a good market quality carcass? Are you selling wool as your primary product and, if so, any particular type? You need to know your market. To use our farm as an example, we hope to earn a profit from our operation. To achieve this we rely heavily on our wool sales to the handspinning market. Within the Corriedale breed we tend to sell our finest fleeces fastest regardless of color. Over the years we will have more of a demand for certain colors and shades than others. We hope that our fiber customers associate us with Corriedale style wool of consistent quality. The sale of our breeding stock also hinges on this consistency. Lastly, in

48

order to more than break even, we must also be able to produce a growthy lamb that (if we choose not to keep it as breeding stock) will bring a good market price. Thus we are selecting for a sheep which produces a specific fineness in the wool, which produces certain colors and shades and which can also produce and raise marketable lambs for the slaughter market.

I believe that a growing concern in marketing wool, breeding stock and slaughter lambs relates to health and environmental concerns. Being able to guarantee that your sheep are free of certain diseases will only continue to grow in importance. A closed flock is a method of controlling a flock's health problems. The regular importation of stock from outside your farm increases the "at risk" nature of your flock, more so if the imports come from flocks with lower health standards than your own. The shepherd who can guarantee that the flock is free of OPP (Ovine Progressive Pneumonia), Johne's disease, foot rot, etc. and who is enrolled in the Scrapie Certification Program may be able to increase the marketability of their stock. The downside to the closed flock is the limitations that it ultimately may lead to in genetic diversity, especially for the smaller flock. To use our farm as an example, we have had a closed flock for over four years and have been able to profit from it as far as sales. We have enough bloodlines with enough rams to keep the genetic diversity alive for a few more years. Ultimately, however, we will have to dip from outside our genetic pool. At that point we must choose between artificial insemination and bringing in new rams from outside. Prior to closing the flock, we had to select our replacement ewes and rams based partly upon their ability to provide us with as much genetic diversity as possible before we closed the flock.

Personal Goals

If you wish to emphasize personal goals with your flock selection, it may work together with the goal of profitability, it may work against it or, it may supersede it. How important is your flock to your personal well being? Many of us derive pleasures from shepherding that cannot be measured monetarily. Working with sheep can be spiritually satisfying. To spend an hour in the quiet early morning on fresh pasture with your flock can provide great inner peace and contentment. Personal fulfillment may lead you to keeping a smaller ewe lamb, who will take an extra year before she can be bred, but who has the personality of an angel and who is able to warm your heart when human contact may fail you. Our farm currently "justifies" a retirement community of older ewes, who have "paid their dues" in both lamb and wool production and who will get to live out their days with us as reward. Their continued presence on the farm may conflict with our financial goals, but the income from their annual wool clip will pay their feed bill. Their continued value is in their personal friendship. Mean or aggressive sheep may not fit into a setting such as ours, even if they have beautiful wool and raise triplets on their own. We have difficulty tolerating a wild or spooky sheep that is not easy to work with or handle.

Genetic Preservation

Lastly, I believe that the shepherd raising a rare breed has the special obligation to preserve the best of the breed rather than to merely increase its numbers without concern for quality. As such, it is especially important that only sheep that meet these standards should be retained and/or sold as breeding stock. This same obligation holds true for shepherds of colored sheep from "non-rare" breeds. Phil Sponenberg eloquently addressed the issue in his article "Colored Purebred Sheep - Why They are Important" in the Fall 1998 issue of <u>Black Sheep Newsletter</u>[4]. My concern here is for the diversity of the genetic pool of purebred colored sheep. Using purebred colored Corriedales as an example, there may actually be fewer in number in North America than some of the rare breeds. With the current continued decline in numbers of large white purebred flocks, the sources of new occasional colored animals decline. In addition, there are no breed standards, as such, for many of the major breeds when they are <u>colored</u>. As an example, colored Corriedales are not recognized by any national breed association. For the breeder of purebred colored sheep of any breed, it is vital not to dilute the genetic purity of your flock whether or not it is a rare breed.

It should be apparent by now that we have selection criteria on our farm that at times work against each other and at other times complement. It should be obvious that our standards for selection reflect our unique situation and are not therefore universally transferable. Now that I have gotten this far, I am not sure whether I have answered the call for a dialogue, but I hope it will encourage further discussion. In the meantime I am off to the barn to discuss the issue with the flock. I just hope they are in agreement when lambing time rolls around.

Happy lambing!

---

[4] Sponenberg, Phil, Black Sheep Newsletter, Fall 1998. Black Sheep Press, Scappoose, OR

Technology and the Old Ways

We have just finished shearing the flock and within the next day or so we should see the birth of our first lambs for the year. As I await the first ewe's labor, I have been marveling at the rapid evolution of farm technology during the last 50 years. Technology has drastically changed the nature of farming and the rural/agrarian society that has depended on that technology. Without becoming nostalgic for "the olds ways" I feel that, due to these changes, we have lost much of what is good for us as a society. The decline of the small family farm and the sense of community that surrounded it will be hard to replace. I wish that in my lifetime I had been able to experience the sense of community that came with such events as threshing or barn raisings.

How did this discussion jump from shearing and lambing to pondering modern agricultural technology? I got started on this line of thought a couple of weeks ago when we were preparing for shearing our flock. Aside from the advent of electric powered shearing machines, technology has not advanced nearly as rapidly into the wool side of farming as it has elsewhere. As a result it is still a labor-intensive job. We are lucky enough in our rural area to have just enough sheep and goat raising friends that we can enjoy some of the community that much of farming in this area has otherwise lost. When shearing time arrives, we often end up sharing our labor at each other's farms. Our shearer, Dave, is excellent -- a true asset to all of our enterprises. In order to justify the long, late winter cross-state trip he must make for us, we have, as a group tried to coordinate our shearing days so that Dave can finish all of our flocks in either one or two trips. In order to make each farm's shearing go smoothly, we share our labor: catching, undressing jacketed sheep, sweeping, skirting and bagging fleeces, etc. At any given farm our group may be a different number or mix of individuals. A lot depends on how many sheep or goats need shearing at a particular farm, along with which shepherds might be nearest to lambing. At each farm there is a lot of work to be done, but a great sense of community always develops. Stories and jokes may evolve as the sheep are shorn or we may talk of technical details of some facet of the particular farm's operation. Without the help of our fellow shepherds we would not be able to do the quality job that the wool from each of our flocks demands. We certainly could not get the job done as quickly, which is a plus for shearer and pregnant ewe alike! In a very small way, it is an experience that allows me to sense the hard work, camaraderie and community that events like threshing or barn raising would have fostered not so long ago. Should we ever have to give up this group effort it would be a personal loss for us; our small sheep community would be poorer for it. Thank you Dave and all our shepherd friends!

Thinking along the same type of lines, I also marvel at the influence of computer technology on the very old occupation of shepherding, even on a small farm such as ours. It is both a blessing and a curse. Much of my record keeping is enhanced with a computer. Gone are the long hand written computations of comparative growth rates of lambs, wool yields and cost/benefit calculations. Just a little time at a computer can yield so much valuable information about the flock, family groups or individuals that, at times, it is nearly overwhelming. It indeed does have an extremely useful place within our operation.

Having said all this in praise of the computer, why is it that I am still drawn to and dependent upon a stack of handwritten notes? Over the course of my shepherding experience I have developed a single set of handwritten notes which, over time, has become very much an integral part of our operation. It is nothing complex or technically demanding and it is probably practiced in one variation or another in many flocks.

During the course of a year running from lambing through the next shearing I will keep chronological notes in a spiral steno pad notebook. The notes will relate to anything having to do with the flock, from the recording of turnout onto spring pasture, to seeding rates for pasture planting, to notes concerning individual sheep. Each year the bulk of the recording occurs at lambing time. The notebook accompanies us to the barn and will record (as time and the ewes allow) any pertinent details about a ewe, her delivery and her lambs at birth. It will also collect entries for illnesses or injuries and their subsequent treatment. It becomes the source of much of the statistical information that is entered into our computer (birth dates, weights, etc.). However, its true value surpasses these quantifiable bits of information. It in effect becomes a diary of our year with the flock.

As it has evolved, the notebook becomes the main source for transcribing anecdotal records for each member of the flock. Usually in the quiet time, about a month prior to lambing, I will take the last year's entries from the chronological notebook and transcribe whatever seems relevant to an individual sheep's record. When the transcription is complete we have anecdotal notes for each member of our flock. The act of transcribing just prior to lambing also serves to refresh our memories about which ewes might have a problem or idiosyncrasy to watch for during the upcoming lambing. When viewed over time the transcribed records reveal much about the performance of a ewe or ram. As often as not, the ewe with the shortest, briefest of histories is probably one of our best ewes, since she has usually had such smooth lambings, problem free lambs and few health problems of her own. Her year's history may be as short as: "During evening chores, Gayle went into labor, delivering unassisted two ram lambs in short order. All three do well. Why can't all deliveries be like this?"

The act of transcribing also reveals a lot about ourselves and our emotional ups and downs, especially during the intense time of lambing. It is easy to see fatigue setting in over the three or four weeks of lambing. Entries become shorter, the handwriting poorer, sentence structure garbled. In short, the author is getting punchy! For example: "What observant shepherds we are. While preparing to dock and castrate Glynnis' ram lamb, we discover that the ram has become a ewe!" or "The lamb is just dumb and his brother is dumber!" The act of recording our experiences with the sheep becomes a bit of an emotional safety valve.

The computer has its place, but it does not travel to and from the barn at odd hours of the day or night in all types of weather. It does not get stained with birth fluids or iodine. It is not with the ewe when she gives birth, nor with the lamb as it struggles with an illness. It is not with us when we grieve the death of a trusted ewe or rejoice in the beauty of newborn lambs.

Even the pen and paper do not venture out into the darkness of 2 o'clock in the morning to wonder at the brilliance of the stars in a moonless sky and to refresh our spirits after many hours with the ewes. Nevertheless the pen and paper still have a place for us in this technical age.

Rambling Thoughts While Waiting to Cut Hay

E arly June is usually such a busy time on our farm that I rarely remember that the <u>Black Sheep Newsletter</u> has a deadline until it is passed. April, May and June for anyone farming in the upper Midwest is always a hectic time. The seasons transition so rapidly from late winter to early summer that one hardly knows where May disappeared. In our case this year we have tried to simplify our schedule by eliminating some of our endeavors. We decided last fall that our sheep were more important to us in terms of labor and income than was our cherry orchard. So after 17 years the orchard is gone and we are out of the business of trying to grow cherries. The orchard has now become more acreage devoted to pasture for the sheep. The only visible clue that the orchard was there is the seven-foot, high tensile electric fence intended to exclude the deer from the opportunity to do wholesale pruning on the cherry trees. It is one heck of a fence for a flock of sheep however! I am still waiting for someone to ask how high a fence our ewes are capable of jumping.

In many ways, a productive orchard and a flock of sheep can be very compatible and complementary. We were able to periodically use the sheep to graze the grasses and weeds that would grow beneath the trees. The sheep would also clean up the fallen leaves in the fall, thereby limiting the over-wintering host source for a fungus that attacks the new growth on the trees in spring and summer. If allowed to, the sheep would clean the orchard floor very effectively and thus save us the task of mowing. The sheep in turn would provide extra fertilization, both directly and indirectly when I spread composted winter manure/bedding in the orchard. There were trade-offs in terms of labor. The sheep would have to be fenced away from the trees, since they loved to prune them as much or more than the deer. Thus their grazing paddocks within the orchard were long, narrow strips, which entailed extensive movement of portable electric fencing daily or twice daily. We could not always allow the flock to graze the orchard at the time that the grasses were at or near their nutritional peak, because of the need to spray the trees for leaf fungus or because the harvest was going on. The sheep did not assist with the harvest in any way except to distract the pickers.

If an orchardist is able to maintain an organic operation, sheep are an ideal compliment. In our situation we attempted to limit the chemical usage as much as possible. Due to certain cherry disease problems in our area it is not possible to be organic with the trees and still keep them alive and productive. There were, therefore, periods in which fungicides had to be applied to the trees. When that occurred, the flock would have to be excluded until the grass could be mowed and washed by the rain. Ours was the first orchard planted on this ground, so we were aware of exactly what types of chemicals have been applied and if they would have presented a risk to the sheep. I would have felt much differently had the orchard been well established prior to our purchase. In the not too distant past cherries and apples were subjected to a barrage of chemicals, some of which remain in the orchard soil years after the orchard may have been removed. Lead arsenic and copper are two especially long lasting residues that may present problems for a grazing sheep. The risk of chemical intake to any animal grazing pasture on such ground can be great. A new owner of an old orchard who wishes to graze his sheep beneath the trees would be well advised to have the soils and forage tested for chemical residue prior to grazing.

I will definitely not miss the hours of annual pruning in the spring or the intense period of harvesting for two weeks in midsummer (at a time that usually coincided with the second cutting of hay!). Our orchard was too small to be able to justify the contracting of a mechanical picking crew. Cherry picking by hand is arduous and tedious work that must be completed rapidly before the fruit is over-ripe. On the other hand, I already miss the sight of the trees in full bloom on a sunny day in late May. It was a picture that was only improved by the presence of the ewes and their young lambs eagerly grazing beneath the trees in the spring splendor. The sheep are now still grazing in the same location, now more frequently than before, which is a bonus, as the "orchard" pasture is adjacent to the house. It affords us more opportunity to just watch the sheep as they graze.

While the departure from cherry production was intended to be a labor saving measure for us, we also felt that we could produce more income from the acreage if it was intensively managed pasture devoted exclusively to the sheep. Of course we also replaced one labor-intensive task with another one (which probably proves that we still are on a steep learning curve). We now have sufficient pasture to graze at least 250 sheep and lambs in the summer and still make enough hay to feed the breeding flock for winter and early spring. In a very good year we even hope to grow sufficient hay to be able to market some of it. In this case we have exchanged a large bill for purchasing hay for a large "labor bill" for baling all of our own hay.

All this brings me, in a round about fashion, as to why I am able to write these ramblings in the height of haying season. Rather than enduring a disastrous drought that was in full flower a couple of months ago and predicted to get worse, it is raining like it did in the "good old days". So instead of cutting, raking, baling, loading and unloading hay, I am waiting for a window of opportunity to just be able to get a few more acres cut in the hopes that it will not get wet. While I wait I can write.

The sheep really put us to shame when it comes to the efficient harvesting and processing of hay. The ewes and lambs are keeping up (barely) with the rapidly growing hay in their pastures. A little rain hardly slows their grazing rate. While I have not even gotten five acres cut in over a week, the flock has steadily mowed across an equivalent field in the same time. The forage has been clipped and is already starting to re-grow. In addition the sheep do not even need to have their oil changed or their joints greased. As a bonus, they have managed a good fertilization program.

Life has been particularly good to the flock this spring. Most of the lambs were born within a very tight window of three weeks in March. They are thus all roughly at the same growth stage. It makes life so much easier for them and their mothers when they are all virtually the same age. It certainly aids us as shepherds with our lamb related tasks. The lambs were introduced to pastures, electric fences and grazing at a point in their lives when they were ready to begin serious grazing. Learning about electric fences and moving to new pasture everyday seems to go more smoothly when the lambs are also yearning for fresh grass as much as the ewes. Thanks to the rains, the pastures have remained lush enough that there has been plenty of forage for the ewes to both produce milk for the lambs and to slowly rebuild their own body condition.

Now all I need to figure out is how to direct the rain to the pastures that have recently been grazed or cut and baled, while at the same timekeeping the sun shining on the fields that still need to be cut and baled. If it keeps raining a bit every day, I should have lots of time to try to figure that one out. More food for rambling thoughts....

Reclaiming the Orchard

Anyone who has read my previous ramblings probably by now has noticed that I am heavily committed to grazing and pasture management as an ideal means of providing a healthy and nutritious environment for our sheep. I would like to revisit our move from a productive cherry orchard to a productive (hopefully) sheep pasture and then put that conversion into the grazing context.

Previously, I related our experiences using our cherry orchard as a backup grazing area. I described our decision last fall to remove our trees and convert the vacated five acres into an additional permanent pasture for the flock. With summer nearly gone, I can now reflect on this transition.

Due to an earlier than normal spring we barely managed to clean up the remains of the approximately 350 trees before the grass began to grow seriously. Grow it did, aided by a nearly steady pattern of alternating days of sun and rain. Since the acreage had been managed for the good of the cherry trees, as opposed to the good of the grasses and legumes which grew beneath them, we did not expect to have the most productive of pastures in the first year. It had been our hope to inter-seed some clover and trefoil using a technique known as "frost seeding".

Frost seeding can only be used in an area where the ground freezes hard over winter. At some point in late winter/early spring as the ground begins to warm, the soil surface will go through a period of alternating thawing and freezing. This process will open small cracks in the soil surface, ideal little germination environments for the legume seeds. To succeed, the seed must be broadcast across the surface of the field at the time when this expansion/contraction is occurring. The relatively large size and weight of legume seeds lend themselves to this technique. Grass seeds, which tend to be lighter, are much less suitable for this method of planting, but they can be used on occasion.

We have used frost seeding as a technique with some good success in past years. We have used it to add to a pasture's plant mix or to rejuvenate the existing stand (especially the legumes) in a pasture. This spring we never got the opportunity to try it. Spring arrived in such a headlong rush that we had not finished removing dead cherry trees from the pasture before the ground was permanently thawed. By late May, the pasture (which we will always know as "the orchard") was ready to be grazed. It was not the most lush, most dense or most diverse plantings one could wish for, but it was deep enough that grazing was in order. It was obvious that there was a need for clovers in the pasture. At that time the pasture was dominated by whatever grasses had established themselves over the previous 18 years. Based upon grazing we had done there in the past, there was a significant population of grasses which were obviously not high on the sheep palatability listing.

By choice, we are not a highly mechanized farm. Our planting options were disappearing quickly since the frost had left the ground. We were aware of one other method for successfully adding legumes to the pasture in its first season of grazing. The method we used has been referred to by a fellow Wisconsin grazier as "Mother's Day Seeding". In Wisconsin we usually experience frequent and plentiful rains in mid to late May, in the period that usually surrounds Mother's Day. By broadcasting seed on a pasture just before a period of rain and then grazing it during the rain or a day or two after, the hooves of the sheep will work the seeds into the damp, warming soil, hopefully making for an ideal germination environment. It is important to have a heavy stocking rate for a short time on these pastures, so that as much of the seed as possible will be trampled into the soil. (Sheep hooves are ideally designed for this task, much better than those of a cow.)

We began grazing "the orchard" on May 23rd, a couple of days prior to a period of rain. I broadcast the clover seeds with a hand-crank seeder that is slung over my shoulder. I covered about 2.5 acres each time I did it. The seeded acreage was always the area that I planned to graze in sections over the next five days. The five acres of "the orchard" were thus subdivided into sections of approximately .5 acres. Each day over the ensuing next ten days, one of these sections would be given to about 80+ ewes and 100+ six to eight week old lambs. In the same period, a good rain occurred about once every third day. By June 2nd "the orchard" had been heavily grazed and nicely trampled. It was now only a question of whether the seeding took and, if so, how well it did.

Within two weeks there were signs of new clovers appearing in the pasture. Was it due to our efforts or was it the emergence of plants from a seed bank already in "the orchard"? By mid July it was apparent that the clover was prospering. It was also apparent that it was the clover that we had broadcast and that the sheep had planted. This conclusion was obvious because the edges of the pasture were scalloped by semicircles of clover, which corresponded to my pattern of walking and turning around with the broadcast seeder. We have had a summer with abundant rainfall and comfortable temperatures, ideal conditions for lush pasture growth. For once our timing had been perfect. So frequent has been the rain that successfully cutting and baling hay has been a challenge. It was three months ago when I wrote a similar statement and we have continued to be challenged right into September. The wonderful aspect of rotational grazing is that the weather has not slowed the sheep with their grazing. By this time in mid-September the ewes and lambs will have cumulatively grazed "the orchard" for about 25 days this summer. As I write the orchard pasture looks better than ever, with a nice dense stand of a mix of clovers and grasses. We should get at least one more extensive graze of about ten days in "the orchard" before a good killing frost slows down its growth for the year.

I write about this method of pasture seeding and improvement because it can be much more widely applied than frost seeding. One need not farm in an area where the ground freezes solid. No heavy investment in equipment is required. There are three basic ingredients for it to succeed. The first is being able to time the seeding to coincide with periods of rain sufficient to germinate and sustain the newly germinated seedlings. Without the rains and moist soil the chances of success are significantly lessened. Second is having sufficient number of grazing animals for the area you choose to successfully work the seed into the soil. Too few animals per acre and sufficient seed to soil contact will not be achieved. Lastly, it really helps to be lucky every once in a while.

Making and Renewing Acquaintances

On our farm, mid-winter is the time when we reacquaint ourselves with individual members of our flock. Despite the fact that the flock now numbers over one hundred ewes, we know each by name and personality. By early summer we have a good rapport with their lambs as well. We spend considerable time with them throughout the grazing season. During that time the flock will walk with me every morning from the barn to their new grazing paddock. The route is reversed every evening at sunset when we bring everyone back home. Once back to the barn there is usually time for "conversation" and more than enough chin rubbing and scratching behind the ears.

Quite often visitors to the farm will want to meet the "girls" and their lambs. So, usually in mid-morning, there will be a small parade of people out to the pastures. By this point in the day many of the ewes will have gotten their first fill of forage and will have time to visit, bringing their lambs with them to check out the new humans. The lambs often become the main attraction of this visit and are quick to learn that they can get all sorts of attention. The visit serves many purposes. It is often the first chance many of our guests have had to meet a sheep up close. (The learning curve for the humans at this stage is often much steeper than for the sheep!) The visit is also an opportunity for Gretchen and me to spend some quality observation time. If there is a change in the group dynamics of the flock, it is often a sign that we need to watch for a ewe or lamb that may have a problem. We may also notice the need for a jacket change, due to repair or need for a larger size. These midmorning visits will normally produce a few "regulars": ewes and lambs who are extroverts who always come for a visit. There is a larger group from which some individuals may decide on any given day to check out the humans. Then there is a smaller group who could care little for us being there (there are more important matters like grazing, chewing cud or sleeping). This later group will never voluntarily find time for us or visitors.

This entire routine changes once we enter the breeding season. This past October we started out with ten different groups of ewes, each with their own ram, scattered across the farm in individual paddocks. This is the only time of the year in which the majority of the flock does not return to the barn overnight. It would be impossible to keep each group separate in the barn each night. For five weeks we pray that the electric fences are good enough to protect each group overnight. We are concerned by the presence of predators (primarily coyotes and neighborhood dogs). It is also the rut for the local deer population; the amount and frequency for deer movement through the pastures at night increases significantly. One or more deer crashing through a pasture at night can cause untold chaos among the sheep.

By this time it is mid-fall and fast moving into winter. The number of human visitors to the farm declines. The presence of a ram in each group dictates that we not take visitors to any of the groups. A ram can be a dangerous, aggressive animal, especially in charge of a group of ewes many of whom are in heat. So for the duration of breeding, the two of us are the sole human contact for the sheep. We will visit each group, first thing in the morning and just before sunset, to check to see who may have been marked since our previous visit and to make sure that all is well with each group. The presence of a ram changes the dynamics of each group. We are watchful of the ram for our own security and to not threaten his dominance of his little harem.

As we enter the heart of winter all of these dynamics change again. The ewes and rams are again separated and are all back in the barn. The pastures are well grazed and often blanketed with snow. If weather and snow depth allows, the ewes have access to a pasture adjacent to the barn. The "boys" have separate quarters in "The Palace", our addition to the old dairy barn.

At this time, we get to reacquaint ourselves with members of the flock. The ewes are entering their second trimester of pregnancy. The flock is dependent upon us for hay and a bit of grain. The entire social interaction between the flock and us changes. Rarely do they have visits from strangers. The flock is "stuck" with the two of us for their entertainment and human companionship. Perhaps the progress of their pregnancies is mellowing many of them. Beyond the fact that we are the bringers of food, we are also the providers of chin and back rubs, ear scratches and conversation. In this changed environment we begin to meet with a different selection of ewes than we did throughout the grazing season. There are still some of the regulars from the pastures. Fuzzie Bear is still the "official greeter". Blondie is the matriarch who rules the entire ewe flock. She is always early to trot over to make sure that "her girls" are being properly attended to. Having evaluated our performance she then melts into the group. There is a gaggle of ewe lambs who have carried over their outgoing personalities from the summer and who will always crowd for attention.

In the winter environment there are additions to the equation. Ewe lambs that were standoffish throughout the time they were on pasture may decide at a specific moment, for no apparent reason, to walk over to get acquainted. Linsey-Woolsey was the first lamb born last spring. She steadfastly refused to have anything to do with us until late October when she decided that it was time to have her chin scratched. Some make the transition slowly. Lou Ewe has only just crossed over the same gulf of apprehension and has still not quite decided if it is safe to be on the other side. On occasion an adult will unexpectedly decide to cross the same gulf. Jerusha, now working on her third lambing, quite suddenly decided to get to know the shepherds. While we value the trust these ewes put in us, it also has a very practical side. A trusting ewe is much easier to deal with should she need assistance when lambing or if she experiences a health problem.

I am, perhaps, ascribing anthropomorphic traits to our flock. Nevertheless, it is an excellent way for us to document and notice changes in behavior. Sheep are much smarter than they are often given credit. This intelligence is what fuels a lot of the behavior we observe. This winter the rams provided ample proof of this theory. We had seen one or more of them "playing" with the latches between their pens. On occasion they have managed to bump a latch enough to open a gate. In the past all that has occurred is that one group of rams has joined the rest of the rams for a bit of head butting and shared hay to breakup an otherwise quiet afternoon. Our solution: an extra rope loop that slides over the latches on all of their gates. The loop has worked just fine, that is until we left the flock alone for a rare evening out (with other shepherds no less!). Upon returning late in the evening we found that all the adult rams had exited their pen in "The Palace" and found their way into the main barn. Three of the group had managed to wend their way through a labyrinth that ultimately gave them access to the ewes. A grand old time had obviously been going on by the time we returned! When we managed to get everyone sorted out and back into their respective quarters, it became obvious that at least one ram had figured out how to slide the rope loop to the side and then lift the latch and push open the gate (a feat that has challenged more than a few humans!). Now the loop is held in place by a second loop that opens only with a fastener. At present the "lads" have yet to figure this one out... but they are working at it. In the meantime, Jael, the dominant ram, now will take any chance he can to find the gate open or at least minimally latched. He is subtle, cruising nonchalantly past the gate, but very alert to its current status. It is behavior well within his personality. He is cool, often aloof, not outwardly threatening, but always making sure he is managing his role as the #1 ram.

Perhaps we feel that we are reacquainting ourselves with the flock at this time of year because we also can spend more quality time with them when they are all much closer as a group. I would like to think that this is also the perception of the sheep toward us. No one illustrated this better this winter than Iowa, a four-year-old ewe. She is generally trusting, but tends to stay on the edge of the overtly social circle of ewes. She rarely ventures forward to solicit attention but will accept a good scratch under the chin if we happen to meet somewhere in the barn. Recently we noticed her significantly favoring a front leg. She was not too happy with the idea of us examining it. We could find nothing apparently wrong, but cleaned the hoof and trimmed it. The next morning when we arrived in the barn, Iowa was uncharacteristically one of the first to trot up to us. The limp was gone. Her visit appeared to be a conscious demonstration on her part of her improved health and perhaps an expression of an ovine "thank you". Since then she has melted back into the edges of the social circle. It seems that she knows us pretty well, much better than we know her. This is why we value the renewing of our acquaintances.

Using Restraint While Lambing

Lambing season at the farm has again come and gone. As always seems to be the case, lambing produced pleasant surprises and some minor disappointments. As this is being written, the lambs are already at least two to three months of age. Within another three months many of the lambs will have been sold. Based upon our current position on the "sheep year" calendar, we are at present making our initial and subsequent appraisals as to which lambs will stay here and who will go. As we have been giving the lambs their "baby" shots (Clostridia and Tetanus shots), we have continued the evaluation process that we started shortly after their birth. Who is growing well? Who looks to have the best fleeces? Which fleeces are the color and/or the fineness that we desire? Whose fleeces will be the most marketable? Who will make the most desirable additions as replacements in our breeding stock? This year the selection process has been simplified over some year's, since the lamb crop was over 62% males. "Why couldn't you have been a ewe?!" is an all too common comment heard in the pastures and the barn this summer. In our operation there is not much room for many males. If there are potentially outstanding ram lambs we may end up retaining one or two. The wethers (neutered males) may have wonderful fleeces but they will only produce income from their wool. The ewes will usually double that income through the production of future lambs in addition to their wool, all the while their food and housing costs are roughly the same as for a comparable number of wethers.

Besides the desirable physical attributes that we look for in our lambs, we also tend to give strong importance to the lamb's family history. Does the lamb come from a family with a history of multiple births? If so, have the ewes been able to put good growth on their twins or triplets? Is the dam a good mother? Does she rarely need assistance with her deliveries? Does she have new born lambs that are up and nursing quickly? Does she take good care of her newborns and does she continue that devotion to them throughout their youth? Most of these traits are inherited. If the family does not measure up to those standards, there are few reasons for perpetuating that family line.

Throughout our brief lives as shepherds, we have always placed great importance on the above mothering abilities of a ewe when deciding whether to keep her and her offspring. A ewe that refuses a lamb at birth is assured a quick one-way ticket to the sale barn. However, having just made this statement, we acknowledge that even though our yearling ewes are expected to deliver and accept their first lambs on or near their first birthday, we experience an occasionally bewildered first-time mother. She will not know exactly what to do with the wiggling, wet creature that suddenly appeared behind her at the same time that she experienced new, strange feelings and sensations. The five to ten minutes that this ewe may spend very tentatively investigating her newborn can seem to be an eternity in the cold early hours of a March predawn. It is an extreme relief when all of the mothering programs finally start running in her confused head. It also makes us all the more appreciative of the first time mother who reacts to her first lamb as if she had delivered many in the past.

On much rarer occasions the ewe's programming is either severely scrambled or non-existent. Every couple of years we will have a first time ewe that will have absolutely nothing to do with her newborn, despite all the tricks that we try on her. This is a family line which is "shut down" rapidly if the ewe does not get with the program. One thing that we have attempted to do with such an unenthusiastic mother is to restrain her in the hopes that she will eventually get the idea that she should raise the lamb that she produced, rather than butting it as if it threatened her very existence.

Our early restraining efforts consisted simply of a rope halter. The ewe would be tied to a jug fence with a very short amount of slack, only enough that it would allow her sufficient room to eat the food and water that was directly in front of her. Usually this left enough room for the back end of the ewe to be bouncing all over the jug (and her lamb). It would still be difficult for a lamb to nurse from this ewe who was resisting motherhood. Thus, our next modification was to sandwich her between some open home-made panels (made for other purposes) that allowed the lamb access to her udder but did not permit the mother to dance around too much as the lamb attempted to nurse.

Over time, we have had moderate enough success with this method of forced bonding to think about building a special pen for the purpose, rather than having to rush around scraping things together after the fact. As usual we never got around to building such a pen when we did not need it. Eventually we decided to purchase a commercially manufactured stanchion, designed specifically for the function. We found a model which would fit perfectly across the front of one of our standard lambing jugs, leaving a two foot area in front of the restrained ewe for feed and a water bucket. Once having overcome our frugality, our headgate sat idle throughout the next lambing after its purchase! It was almost as if the ewes knew what we had purchased and did not want to be caught dead in one.

This year's lambing also started out as though our investment would not have paid for any savings in lamb milk replacer. However, as we entered the last two weeks of lambing it was finally put to the test. In a period of 12 days we experienced four different yearling ewes that refused to willingly accept their newborn lambs. Our long neglected purchase finally got a workout that we had never anticipated. Thankfully, none of the four ewes lambed on the same day. Luscious, the first of the four ewes, had a difficult unassisted delivery of a large lamb, which once delivered she promptly and totally ignored. Once "imprisoned", Luscious allowed her lamb to nurse, first grudgingly and then willingly. She spent a little over three days in the stanchion, a period of time that we had read was usually sufficient to effectively bond the two. By the end of the third day, the lamb was sleeping under her mom's chin. When we removed the headgate, Luscious and her lamb got along as if there had never been a problem.

Five days later the same scenario repeated itself. Lilac also had a difficult unassisted delivery of a large lamb. Never once throughout her delivery did Lilac lie down, which is not that common with our ewes. Once the lamb hit the ground, Lilac appeared shell-shocked. When we sought to move the lamb in front of her she would repeatedly butt it. So...we set up the headgate again. She was even less pleased with it than Luscious had been and only grudgingly let the lamb nurse. Two days later almost the same situation occurred with Lou Ewe and her newborn lamb. We were nervous that Lilac had not yet bonded well to her lamb, but now we needed the stanchion for Lou Ewe. Once released Lilac was still unsettled (she desperate wanted out of the jug, but she tolerated the lamb). In the headgate, Lou Ewe was as unhappy as her predecessors, but the lamb managed to nurse and slowly the bonding process again began.

Finally within three days of Lou Ewe's delivery, Lisalotte gave birth to her first lamb. It was a delivery that we slept through. (We try to be there for all of the births but toward the end of lambing it becomes more difficult, and sleep becomes more valued for us.) Somehow Lisalotte's lamb was squeaky clean when we found it, indicating that mom was initially very devoted to it. But it was now across the barn from Lisalotte, who now would have nothing to do with him. Lou Ewe gained an early release; she and her lamb were well bonded and devoted to each other. Lisalotte took over the headgate. It took her much longer to "get religion". After three days, she still refused the lamb when we released her, so we put her back into the stanchion. After another two days, she too came around to the idea of motherhood.

So, how have these four mothers and their lambs done so far? All four ewes remain well bonded to their lambs. The growth rate for each of the lambs is as good as or better than that of the other lambs from this year's yearling ewes. Had we not had the headgate I am sure that we would not have been able to salvage at least two (or perhaps more) of these relationships. The headgate has therefore already paid for itself in the cost saved on milk replacer and time we would have spent artificially raising the lambs. The growth rate of these lambs is also better than had we to raise them as bummers.

The greatest test will come next spring with these same ewes and their second lambing. At this point we do not know if we have retained ewes that will again repeat this type of performance with their newborn lambs. It will be difficult to tell if they produced lambs with similar inclinations since all but one of their lambs was a male. Despite their shaky starts as mothers the four ewes possess other admirable traits that we would like to perpetuate. In all four cases, the ewes have lovely, very marketable fleeces. They all come from good family lines. Only in the case of Luscious' grand dam has there ever been a similar performance. The other three ewes come from long lines of excellent mothers. Stick around; we should know some answers in a year. In the meantime, the headgate will be stored in a clean, well-protected spot, a place that it has so ably earned.

Communication

One of the great values of a publication like the <u>Black Sheep Newsletter</u> is its ability to share thoughts and ideas among a far-flung community. In the case of <u>BSN</u>, this community centers around fiber producing animals and the use of their fiber. The dialogue generated in <u>BSN</u> has now continued for over twenty-five years and has, hopefully, been a benefit to the evolving group of readers and contributors throughout that entire time. While I have been amongst those readers for much of that time, I have only actively contributed for a shorter while. I often find it difficult to be a contributing writer on a regular basis because I am not, by nature, a writer, nor do I consider myself to be an expert on most of the subjects that I try to address. However, over the last few months, I have become more cognizant of the value of any type of communication within this community.

There is a need for greater one-on-one communication within our sheep/goat/fiber community. Let me offer some examples from recent personal experience. Most of us are involved either in the production, sale or purchase of fiber animals or their fiber. How often have you heard from someone to whom you sold an animal for breeding stock? Have you gotten feedback from your fiber customers about the fleece you sold to them? Have you let a producer know how the ram or ewe you purchased from them is doing? Have you had problems with a fleece you purchased? Have you had a wonderful experience spinning a particular fleece but not mentioned it to the producer? Have you tried contacting the producer for assistance? In all of these scenarios, there is potential for valuable positive information exchanges.

As producers of breeding stock it should be important to us to know how the sheep that we have sold have worked out for the buyer. Nevertheless I am amazed at how often a ram or ewe leaves our farm with its new owner and they seemingly drive off the edge of the earth, never to be heard from again. As a buyer, it is so very important to provide the feedback to the breeder about the performance of the animal. On occasion we will hear about the achievement of an award, e.g. the winning of a fleece competition. It is rarer to learn about the ram's performance. What has been his prolificacy? What kind of fleeces are his offspring producing? Are the lambs robust and growthy? In a couple of situations we also have become aware much later that the buyer has not achieved the results that they wished from breeding that ram to their ewes. Had there been greater and/or more frequent communication many of those difficulties could have been solved. As an example, we have had a couple of situations where a colored ram is purchased to add colored genetics to a flock, but the buyers cannot understand why when they breed that ram to their white ewes that there are never any colored offspring. A quick and simple lesson in recessive color genetics would have solved the buyers "problem". But we may only hear about this situation indirectly or after a number of years have passed since the sale. A timely status report from the buyer would have produced a fairly timely solution to their problem and valuable feedback to us as breeders. On the positive side it is especially gratifying to hear about success. Often receiving those success stories provides us with an emotional warm, fuzzy glow that helps make our efforts seem worthwhile.

Similar experiences can be of value to the fiber producer and the fiber buyer. If you have purchased a fleece that does not do for you what you wanted, you need to let the producer know. In our case we produce enough fleeces each year that we will never be able to personally experience spinning wool from each sheep in our flock. Communication from our customers may be the only way that we learn something special about the wool of a particular sheep. Often what seems to be a dissatisfying condition may be resolved by the producer's knowledge of the specific fleeces and the sheep from which they originated. For example, if you have never purchased a fleece from a particular breed there may be special tricks to washing, carding or spinning that type of fleece. Even though we provide special written instructions with each fleece that we sell, we still occasionally hear from buyers who have processing questions and who have obviously ignored our suggestions. For every person who takes the time to contact the fleece producer with their problem, there are probably still more who do not make the effort to communicate, but just needlessly give up on that producer or breed of sheep for future purchases.

As a producer, it is therefore necessary to make a conscious effort to follow up with your customers to determine if they are happy with your fleeces, if they need some sort of assistance with them, or if you as the producer need to make some changes and/or improvement in your fleeces and their presentation. It is nice to hear from a third party that a person who purchased one of your fleeces is delighted with it; it is much more gratifying to hear that directly from the customer. It could be devastating to find that a customer is complaining about your fleece to everyone else but not to you.

Healing

It seems somehow appropriate as I write in mid-December, that the year 2001 is ending for us in as much of an unsettled manner as has characterized the previous twelve months. The sheep are currently out on pasture contently grazing. They returned to pasture today after having safely survived the dangers of the last few weeks of the deer hunting season. That they are able to be grazing at this date is amazing. Rather than dealing with snow on the ground, they are faced with grass that is still growing, albeit very slowly. We have yet to see a flake of snow and there is no frost in the ground. It is a pity that we do not have significant forage stockpiled in our pastures to allow the sheep to gain the majority of their feed on their own for at least a little while longer.

Mother Nature seems set on putting her strange twist on the end of a year that has seen so many manmade twists to all of our lives. Our outlook on life has changed dramatically since the world events of September 11th. We have a new, deeper appreciation for the people we love and to whom we are close. Our relationship to our flock has also changed for many of the same reasons. Yet, it is not the events of September 11th that are solely responsible. For us, as providers for and protectors of livestock, the events of the world began to close in upon us much earlier.

The outbreak of Foot and Mouth Disease in Great Britain in the early spring of 2001 continues to send its rippling effects across the ocean to our doorstep. The pain and suffering endured by our fellow shepherds and their flocks in Britain is still almost unimaginable. Their losses, financial, physical and emotional, have been extreme. The senseless loss of life for so many healthy sheep, cattle and pigs is staggering. Here at our farm our reactions first were confused and then quickly extreme. Trying to imagine the effect upon our lives if we had to experience similar losses was sobering. What would it be like to have to destroy our entire flock, the labor of eleven years of breeding and love? Our personal connection to each of our sheep was strengthened as we tried to picture life without them. Were we willing to make major sacrifices in our lives in order to minimize what was a very real threat?

Our flock is exposed to a much greater degree of diverse and direct human contact than many flocks in this country. Our other major source of income is from the operation of a traditionally styled bed and breakfast. Over the years, the bed and breakfast, the sheep and their fiber have become mutually supportive of each other. One of the special draws to our bed and breakfast has become the opportunity for our guests to quietly meet most of our flock up close, on a daily basis. In addition, they are able to see and purchase finished fiber items which have a direct named connection to the sheep they have met. Interestingly many of the sheep seem to now look forward to this contact with strangers as much as the guests look forward to meeting the sheep. These daily visits have become a tool for us to preach a gospel of sustainable family farming. For the vast majority of our guests, this morning visit has been their first chance to get anywhere near a farm animal. In a small way, it has become a valuable lesson for many as to where their food and clothing can (and should) come from.

The quite real possibility that our flock could be exposed to Foot and Mouth or other diseases suddenly altered our willingness to share this wonderful encounter with others. Like many, we hung up bio-security signs, warning visitors to stay away from our pastures without our supervision. We became much more hesitant to take our guests (or anyone else) on the "daily tour to meet the sheep". We took to extensively screening potential guests before we would take their reservations. We turned away potential guests from parts of Great Britain, the Netherlands and France, as well as Americans who had recently traveled in those areas. In retrospect we probably overreacted in some ways, but we did have a greater sense of security as a result. We could only hope that others in our immediate farming community felt the need to be as careful.

Throughout the entire experience we were continually amazed by the great degree of concern that our non-farming visitors showed toward our sheep and us. On occasion, we did not even need to screen potential guests for Foot and Mouth risks, as they offered such information as though they had been prompted to ahead of time. On the downside, it also became evident that a number of people were also confusing the Foot and Mouth outbreak with the ongoing BSE/Mad Cow Disease battle. There was still room for much corrective education and explanation, not to mention reassurance. On the whole, however, it was heart warming to hear from so many people how much the health and well-being of our flock also meant to them. It was evident that the tragedy in Great Britain had deeply touched the emotions of many people in this country. They expressed a much warmer and greater appreciation for our sheep and our life style than we had otherwise experienced.

We continued to maintain a heighten level of bio-security throughout the summer and fall, but we slowly worked back into much of our older routine, with guests visits to the sheep. (The sheep appeared appreciative!) Over time, as the seasons progressed, the situation in Great Britain became more distant to most folks with whom we dealt. It had, however, left a mark with many people who stayed with us, an emotion that resurfaced as they met the sheep.

The tragedies of September 11th, 2001 introduced an entirely different set of emotions to much of our world, yet in many ways they were emotions that were extensions of what we had been experiencing throughout the spring and summer. Our farm and the life we share on it with our sheep became all the more precious and fragile. The farm became smaller in relation to the rest of the world, but became much larger for us in our ongoing existence. We came to reassess our life and goals on the farm. Over that time we have made some major decisions regarding the future of the land on which we live, where we hope to go with our sheep and how the farm as a whole figures into our future.

The dependence of the sheep upon us has been reinforced, yet at the same time we have become more emotionally dependent upon them. Our sheep have also proven to be a salve to the emotional wounds of others. People did eventually come back to the bed and breakfast after September 11th. The chance to watch the ewes and their lambs eagerly parading off to a fresh pasture each morning was a small reassurance that there was still happiness and good in the world and perhaps still hope for our more peaceful instincts. Guest visits with the sheep became more prolonged. Greater personal ties to individual sheep and the flock as a whole became more obvious. In their own quiet ways, the sheep were repaying their human visitors for the concern that they had expressed toward the flock during the height of the Foot and Mouth outbreak. They were providing a valuable emotional therapy that the human world could only struggle to provide.

Now, as we enter a winter that is not yet a winter, the guests are largely gone. We have the sheep to ourselves and we value their friendship and devotion more than we have before. As we approach shearing and then lambing this spring we will have a greater reverence for the gifts of wool, lambs and emotional warmth that the flock will present to us. We can now let them devote their energies toward our personal emotional healing.

Postscripts

Once again, what a crazy, confused mix of seasons we are experiencing. It seems so long since we have had a "normal" run of weather that I begin to wonder if we will ever return to those patterns. In Wisconsin, we again had an unusually dry and warm winter. Snow removal and skiing were a novelty! The flock prospered in the mild weather. Even in January, there was still some "green" in the pastures, but very little to graze. We should have known that all of this would not continue.

The first signs of foreboding started in mid-February, when we normally look for a warm "window" in which to shear the flock. The warm window was present, but the shearer was not. It was nearly a miracle that Dave managed to get here by the end of the month. His recovery from an appendix that burst in January was astounding. But as a result, the sheep did get sheared about two weeks later than expected and not a moment too soon, as they began lambing the next week. The earlier foreboding was justified. As soon as shearing was over, the weather turned unseasonably cold and damp. We found ourselves in "deep winter mode" just when things should have been warming up. These conditions were, after all, why we decided not lamb in January or February! We had freshly shorn sheep that needed protection from the cold and dampness. Lambs began arriving in the midst of the cold and damp.

Opportunities for pneumonia and hypothermia in the lambs were ideal. We experienced an abnormally high number of pneumonia cases and lost some lambs that in a "normal" year we would have expected to survive.

The weather had other unanticipated consequences. Once sheared, we found that our fleeces were especially greasy. We thought originally that it might have been a problem for us alone (nutritional perhaps). Yet as other flocks in the state were shorn we heard similar tales from their shepherds. We can only guess that the warmer winter weather stimulated more lanolin production and/or secretion. Our challenge was to try to alert and educate our fleece buyers on the need for extra care and thoroughness in their fleece washing. As it is, Corriedales are always a very noticeably greasy fleeced breed. Good washing techniques and detergents will be especially important this year.

By June, we had barely experienced spring type weather, as it remained cool and wet. While much of the western part of the U.S. cooks in heat and drought, the ground here has been so wet that spring planting has been severely delayed or made impossible. In low lying or clay-based soils, it has been difficult to get grazing animals out on pasture as the ground is so saturated. At our farm we at least benefit from sandy loam, well-drained soils. The sheep and lambs have been on pasture since the first week in May. They are having difficulty keeping up with the pasture growth. Despite the cold temperatures, the rains have kept the grasses growing rapidly. Making hay is going to be a large challenge should this weather persist.

All these aberrant developments have resulted in some lessons for us. What have we hopefully learned? What follows are "postscripts" either to what I may have written previously or to procedures that had changed for the first time in many years due to the different conditions.

Ultrasound

At the eleventh hour, when it seemed that we might not get the sheep shorn before lambing, we nearly panicked. I had not shorn any sizeable number of sheep in about eight years. My skills (what little I ever had) were rusty and my conditioning for the task was non-existent. The last time I sheared, the flock numbered in the 30's, not the 124 that awaited this year. The quality of my work would not be up to our standards and the fleeces would therefore not be as good. As we could not locate a quality shearer on such short notice, we decided that I would at least crutch the ewes if they could not be shorn prior to lambing. Since the task still seemed daunting, we decided to lower the number by only shearing the ewes that were pregnant. Luckily, one of our veterinarians, Gavin, had ultrasound equipment available. As it was, some of the ewes were at the very latest date when it would be a trustworthy technique. As the pregnancies advance, it eventually becomes difficult to identify a fetus using ultrasound, as there is too much amniotic fluid present to get a good reading. When we decided to have the ultrasounds done, our vet advised that some of the ewes might be so far along that he could not promise an accurate test. It would have been much better had we been able to schedule just a few weeks earlier. The testing went smoothly. There were only three ewes (all with early breeding dates) that Gavin could not make a call of "yes" or "no". Once lambing was over, with the exception of those three ewes, his predictions of pregnant or open were 100% accurate.

As it turned out, we did not have to crutch the ewes, so the original reason for ultrasounding them was moot. Nevertheless we did benefit from the knowledge. It permitted us to separate open ewes from pregnant ewes, thus allowing us to not over feed and over condition the open ewes and to provide more liberal space for ewes in labor. If we test before next year's lambing and if shearing is on schedule, it will also permit us to collect a fleece from an 11 month old open ewe and then decide if we wish to immediately market her (thus receiving the significantly higher lamb price instead of old ewe price). The latter would further open up housing possibilities. There are some ultrasound technicians that have equipment that can provide an accurate fetal count and estimate their due date. If such information were available, it would permit separating the ewes with single lambs from those with multiples and then feeding each group accordingly. Planned in advance, it may become an annual procedure for us. It may well justify the extra cost, in terms of better ewe and newborn lamb health.

Stanchions

A year ago I wrote of our experiences with stanchioning four ewe lambs who initially rejected their newborn lambs. We retained each of these ewes, as they eventually did a good job with their first lambs (once they were convinced that they needed to care for them). Each ewe had some nice fleece characteristics that we wished to maintain. This year we approached each of these ewe's delivery with a degree of anxiety, as there is some indication that this behavior may repeat itself and may be hereditary. Now that lambing is over we can say that last year's effort was worth keeping three of the four ewes.

88

Luscious was the first to lamb. She was exceedingly excited when her lamb was delivered. Her enthusiasm for cleaning and caring for her lamb was so boundless that we had the impression that she remembered only too well three days in the headgate a year previously. She and her lamb were tightly bonded to say the least! Lou Ewe and Lilac mirrored similar behavior with the arrival of their lambs (albeit not quite so gushingly enthusiastic). It was especially gratifying in Lilac's case since her stanchioning experience the previous year was much more stressful for her and the lamb. Lisalotte was our only failure. She delivered twins and refused to have anything to do with either of them. Restraining her this year proved pointless; her physical struggle was extreme and even restrained she did everything she could to fight the lambs. In retrospect we perhaps could have foreseen this rejection. As a first time mother she took much longer than the other stanchioned ewes to finally accept her lamb. When she accepted that lamb there appeared to be much less of a permanent and strong bond between them. If we experience a similar lack of bonding in the future with another ewe, we will not give her a second opportunity. Overall, using the headgate was worthwhile. We have managed to salvage three good ewes out of a group of four.

Mother's Day Seeding

Two years ago we had great success renovating a pasture using very little mechanical efforts. It involved manually broadcasting clover and trefoil seeds on pastures immediately prior to predictions of rain. Once the rains came we let the sheep heavily graze the pastures with the hope that they would trample the seeds into the warm wet ground and thus insure a good germination. In our locality mid to late May is an ideal window for this technique (hence the name). Two years ago this seeding was an extreme success. By mid summer it was apparent exactly where I had (and had not) broadcast the seed.

Last year we repeated the process on some other pastures. Our timing was still in mid May and the sheep did intensively graze the bulk of these pastures shortly after good rains. The results were seemingly unsuccessful. No lush stands of clover appeared by mid summer and little was evident in the early fall green-up the pastures experience. It was therefore a real joy to discover clover and to a lesser extent trefoil in great abundance this spring, a year after the seeding. For whatever, reason the conditions were not as ideal for immediate germination and/or for immediate growth. The abundance of rain this spring was the apparent catalyst. So the technique has merit, especially considering the small investment in labor. Timing and technique may be important. Patience is crucial. The same could be said for much of good shepherding.

Artificial Insemination, Part 1: Kicking Tires with the Boys

It is always fascinating to me to observe the different personalities that each year's lambs bring with them to the farm. Even though the underlying genetic base for our flock has not changed for seven years now, each new crop of lambs seems to have their own unique personalities that often do not resemble those of the previous year's lambs. This year's ram lambs are especially noteworthy. We left nine of the ram lambs intact this year. As always, one hopes that they were the right choices and that those that became wethers were correctly chosen to that role. The ram lambs were weaned at about three and one half months of age. (We learned the hard way that leaving an intact four month old ram lamb with our ewes can too often result in the unexpected arrival of newborn lambs in November!) Of this year's group of ram lambs, two are destined to stay with us and the remainder were offered for sale as breeding stock. This year's group seems to have been especially close knit. There has been the usual amount of head butting amongst them, but nothing particularly serious. As a group, they seem to have enjoyed each other's company more than most our groups of ram lambs. They also seem to have a greater sense of curiosity than ram lambs from other years. That curiosity, mixed with a bit of boldness, has led them into more than their share of mischief this summer. That personality has stayed with the group, even as the group has gotten smaller due to the sale of many of them.

Nicely Nicely and his buddies checking out the tractor

The similarity of the group to a comparable number of teenage American boys is at times astounding. I was recently reminded of this similarity as I was getting the tractor out of the shed in preparation for making hay. At the time, the boy's pasture partially surrounded the machine shed. In order to get equipment in or out I either must temporarily herd the boys back into the barn or just try to work around them without letting them escape into the greater outside world. On this occasion it just seemed easier to work around them. Upon opening the shed the ram lambs all ran inside seemingly to check out every piece of equipment in the place. This was the first occasion that they had to meet a tractor up close, even though they had seen it drive by quite often. Even the roar of the tractor starting up did not seem to faze them. After hopping off the tractor to flush the lambs out of the shed, I climbed back on the tractor and drove it outside. No sooner was the tractor outside than the boys came rushing over to get a closer look at the vintage Massey-Ferguson, even with the motor still running! They were the classic group of guys gathered round the new car/truck/tractor, checking out all of its features. If they could have gotten high enough they most assuredly would have wanted to check out the engine. As it was, they were kicking the tires, except that in their case it amounted to butting them! The image of a 120-pound lamb going head to head with a large tractor tire was timeless. This was "American Graffiti" revisited!

Later, I began to wonder if North American ram lambs had different "coming of age" rituals than, for example, ram lambs in Australia or Mongolia. Perhaps next spring I will find the answer, as we are about to attempt to bring some Australia bloodlines into our flock using artificial insemination (AI). However our adventure into AI is not based upon my need to answer that question.

Artificial insemination in sheep is not a new procedure, having been refined since the 1980's. I thought that I would, over the course of the next few months, relate our novice, first time experiences with the AI procedure, in the hope that it will shed some light on the process for those shepherds, like us, who have not had the opportunity to experience or observe the entire process.

For a couple of years we have been kicking around the idea of using AI. Our major reasons were two fold. First, we have maintained a closed flock for over seven years for health security reasons. By not bringing new sheep onto our farm we can drastically limit the introduction of health problems for the flock. As long as there is enough genetic diversity within the flock, a closed flock will not present inbreeding problems. When we closed the flock we had enough diversity to be able to operate for a fairly long time. We, however, have decided not to press our luck too far. Introducing new rams using artificial insemination presents a good solution to most biosecurity issues we face.

The second reason we have been interested in AI has been the possibilities that it presents for introducing blood lines from outside those usually available in this country. Specifically, we were curious to see if we could add to the quality of our Corriedales (especially in regards to their wool) by introducing Australian and/or New Zealand genetics. Those two countries were chosen as they are the original sources of the Corriedale breed and because we do not want to concern ourselves with the introduction of scrapie (as both countries are presumed to be scrapie free).

The first AI lesson that we have already learned, through sometimes difficult experience, is that the artificial insemination procedure requires a good amount of underlined advanced planning. We needed to have everything set up well before the actual date for breeding. Finding the right rams overseas is a challenge. Dealing directly with an overseas breeder of colored Corriedales, who has not exported to the US before, can be daunting. Dealing with numbers (e.g. ram weights, fleece yields, micron tests, expected offspring production, etc.) can have its advantages. It eliminates the intangible, gut reaction one may have to a ram, but it is hard to substitute for at least a hands-on examination. (Among other things we will probably never know if the ram liked to "kick tires" as a youth or if he was more culturally refined!)

Once we found characteristics that we thought would be good for our flock, we still needed to get the semen here for our use. As time began to catch up with us, we decided to continue our direct pursuit of overseas semen, but to target the fall of 2003 as our date for its use. For this our first year of AI experience, we opted to rely on the services of a US importer. He offered a limited, but seemingly good selection of Corriedale rams from Australia and New Zealand, all but one of which was white. We decide to try three Australian rams, two white and one colored. We will concentrate mostly on breeding our white ewes with AI this year and then try to work on the colored flock in 2003. Currently we plan to AI 13 white ewes and 2 colored ewes, a little over 10% of our ewe flock.

Finding an available skilled inseminator can also be difficult. The most successful technique entails a laparoscopic insemination directly into the ewe's uterus, which is a minor surgical procedure. We were lucky in also having a "local" Wisconsin specialist, Randy Gottfredson, available. Randy is part of the University of Wisconsin sheep program and has done the AI work on the university's flock, including its large dairy sheep flock.

Once the semen has been lined up and the AI technician contacted, a breeding date had to be decided upon (a date that one can be positive about scheduling the AI technician!). In our case that date was set for September 25th. We then had to count back from that date in order to start the synchronization process. The ewe's estrus cycle is controlled using hormone therapy. The process is usually achieved by inserting a pessary or sponge impregnated with progestogens into the ewe's vagina. With the proper tool, Gretchen and I found this to be a relatively easy process for the two of us. The increased level of progesterone prevents the ewe from coming into estrus. The pessary is removed 12 to 14 days after it was inserted, at which point progesterone levels drop and follicles develop. To further tighten up the synchronization a shot of PMSG (a special restricted use drug) is given to the ewe at the time the pessary is removed. Insemination is then supposed to take place within 56 to 66 hours.

Like in any normal natural breeding program, the ewe needs to arrive at breeding time in good condition and on a rising plane of nutrition. We are planning our AI breeding to occur about two weeks prior to putting the rams in with the rest of the ewes. Hopefully that will allow us to put our AI ewes in with a clean-up ram about a month after they were AI'd. By then, if they were bred they will have had time to settle and, if they are open, the clean-up ram should be able to take care of then without causing an overly extended period for lambing.

We also had to make good selections for the ewes that we planned to AI. In addition to being in good condition, we looked for ewes that had lambed at least twice and who tended to produce multiple births. Looking at their past lambing records we also looked for ewes that always bred early in the breeding process, i.e. during their first heat cycle. That tendency should be a good indicator of high fertility. They needed to be good mothers, who consistently have been able to raise good, healthy, robust lambs on their own. For our operation, which focuses heavily on quality wool production, we also sought out the best of our fleeces, in the hope that the fleece quality of the AI offspring would be especially good. Ultimately our primarily goal of this program of importing new genetics into the flock is to further refine the quality of our wool.

As I write we have selected our ewes for the artificial insemination and they have received their hormone therapy sponges. In another ten days we will be doing the actual insemination. With that they will hopefully all be bred. Next time I hope to report on the actual breeding event and the subsequent experience. By then we should also know how much these new rams will have cost and, perhaps, how effective they were. We will have to wait until next year, however, before we learn if their offspring are also into kicking the tires on old tractors.

Artificial Insemination, Part 2: The Novice Perspective

When last I wrote, the flock, Gretchen and I were all about to enter an entirely new realm of experiences: artificial insemination (AI) of sheep. The previous year we had the opportunity to observe and "assist" with the insemination of some of our friend Julie's Shetland ewes. That all too brief opportunity constituted our only direct experience with laparoscopic artificial insemination. But it was enough to inspire us to make similar efforts with our flock, especially once we considered the very high success rate for the insemination of Julie's ewes. I will continue to recount our first time, novice experiences with AI, in the hopes that it may be of some aid to other shepherds contemplating its use for the first time. I cannot complete the tale this time since, as I write, our AI ewes are about half way through their pregnancies. Next time I should be able to write a final chapter.

We decided to attempt to artificially inseminate 15 of our ewes this fall, using semen collected from three different Australian Corriedale rams. The three rams were selected based both on their history of heavy wool production and the relative fineness of their fleeces. September 25th was the target date for the insemination. When I last wrote, we had already started the process of synchronizing the ewes' estrus cycles by inserting a sponge impregnated with progestogens into the ewes' vaginas. While we waited the two weeks that the sponges needed to remain in place, we tried to make sure that everything else was in readiness for the actual AI event.

The first task was to insure the timely arrival of the frozen semen. It was shipped in a container filled with liquid nitrogen, which was in turn inside a larger shipping container, designed to protect the smaller container and its contents. We were assured that the container would remain sufficiently cold to protect the semen for at least ten to fourteen days, as long as it was properly closed and no liquid nitrogen was leaking. Five days prior to the date of insemination a large container, the shape of a huge toad stool, was delivered by our regular UPS guy. Randy was used to a lot of strange deliveries to our farm, but this one may have topped them all. When informed that he had just brought us "three studs in a tub", he just shook his head and left!

The tub spent the next five days in a spare bedroom. We are lucky enough to live in an area still relatively strong in dairies, which also means that we have a local cadre of bovine AI technicians. I was able to call upon a neighbor who does lots of bovine AI (and who raises a few sheep on the side). Jim was able to give us pointers on monitoring the liquid nitrogen level in our tank. It was also reassuring to have him offer to store the semen if our tank ran low of liquid nitrogen. On the other hand he told us that sheep AI was much too complicated for him. Luckily we already had our expert technician, Randy Gottfredson, from the University of Wisconsin sheep program, all lined up for the event.

About 56 hours prior to the AI date we were to remove the sponges and then administer a shot of PMSG to each of the selected ewes. The PMSG is a special restricted use drug, supplied by

Randy, which is given to tighten up the synchronization. It is crucial that the AI take place within 56 to 66 hours after the shot is administered. Hence it meant an early wake up call so that we could be down in the barn at 4 AM to give the shots.

By September 25th, we hoped that we (and the ewes) were all set. We had lined up a crew of helpers, all fellow shepherds who are also trusted friends. In addition we scheduled Joel, one of our local large animal vets. (For reasons of liability Randy would not proceed until the vet was satisfied that proper surgical procedures were followed. As it was, Joel stuck around through the entire procedure, largely due to professional interest.) For the previous 24 hours we had withheld food and water from the ewes in order to reduce the content of the bladder and rumen, making it a little easier to surgically get to the uterus. One o'clock arrived followed shortly thereafter by Randy. Once he was set up, the procedure was to go as follows.

Each ewe was placed on her back and strapped into a special cradle that Randy provided. Any wool on her udder and in the area about 6 inches in front of it was clipped as close as possible. This area was then scrubbed with a surgical soap and disinfectant. A local anesthetic was injected into two sites about 4 to 5 inches in front of the udder and 2 inches either side of the middle of her tummy. The cradle was then elevated and tipped so that the ewe's head was much lower than her rump (thus shifting her internal organs toward her head and away from the uterus). Two small incisions were made at the sites of the injections. Trocars, sharp pointed tubes with cannulas inside, were punctured into the abdominal wall. The cannulas (small tubes) were left in the cavity as the trocars were removed. In is through the two cannulas that an endoscope and a probe are inserted. The endoscope is a special lighted viewing tube that the technician can use to see inside the ewe's reproductive tract. The probe is used to manipulate the uterus if it is not properly positioned. Through the scope Randy is able to see if follicles are present in the uterus. If none are present he will not waste the semen on the ewe since fertilization will not occur. He then replaces the probe in the other tube with the inseminating gun, containing the thawed semen. The semen is then injected into the lumen on each side of the two uterine horns. The insemination is now complete, the tubes removed, the ewe lowered to a level position. The two incisions have an antibacterial agent applied and the wounds are closed with michel clips. We lift the ewe out of the cradle and off she goes, headed to a feeder full of nice hay. Within a few minutes she is eating heartily and drinking. Prior to finishing with the semen from each ram, Randy also prefers to examine a small sample to get an idea of its motility and the percentage of viable sperm.

Randy assures me that he and one other person can do the entire procedure, but the time involved is significantly greater with just two involved. Our crew consisted of:

1- the AI technician: Randy,
2- someone to thaw the semen and prepare the straws for the insemination gun: Gretchen the former nurse who received on the job training from Randy,
3- the pre-op crew: Nora and Judy, who did the shearing and scrubbing,
4- the vet: Joel, who made and closed the incisions and then watched over each ewe after surgery (and who provided extra muscle moving the sheep and cradles), and lastly
5- me: the catcher, lifter, gofer, uncoordinated coordinator and worrywart.

When all was completed in just over two hours, we were exhausted. The fatigue was due to our own anxieties over the unknown as much as it was due to the actual labor. Knowing now how the entire process works, I am sure that we could easily cut our time in half the next time we try AI. As it was, we spent more time than was technically needed, as Randy made the effort to teach us as well as inseminate the ewes. We were also relieved that all of the ewes seemed to react well after the surgery. Since then, none have displayed any problems. Not everything went smoothly. One entire set of semen straws (all from one ram) did not cooperate. Even Randy could not get the straws to unplug properly. In the end those seven straws may all have been compromised, as they had to be cut open, thawed and then placed into new straws. One other set of straws was entirely empty and therefore offered nothing with which to breed.

Two weeks after the insemination, we put all of the ewes, including the AI ewes, in with the rams. Within the first week with the rams we felt that we had a good idea of how well the AI went. As we feared, all the ewes that we exposed to the semen from the compromised set of straws were marked by our rams. A major disappointment, since we will not see any offspring from one of the Australian rams. However, of the other eight ewes that we exposed to the other two AI rams, only two were strongly remarked by our rams. Two others received faint marks, hopefully a sign that the ram was just over enthusiastic rather than truly sensing a ewe in heat. Discounting the bad set of straws we are very hopeful for 4 of 8 ewes (50%). If we are optimistic, we may have 6 of 8 ewes (75%). We have been told that a success rate of over 50% is not too bad. It has now been two and a half months since the insemination took place so the ewes are well into their hoped for pregnancies.

Would we or will we do this again? The answer will have to wait until late February 2003 when we will definitely know the outcome of the experiment. From a technical standpoint, we feel that we are now much better trained and prepared for the task. We have made some mistakes and learned from them. If we are to attempt AI again, we should be able to do it better and more efficiently. We will also know a lot of questions that we did not know to ask the first time around. Next I hope to outline our mistakes and successes. Hopefully I can also gush over the wonderful new lambs sired by rams from both sides of the Pacific. Happy lambing to everyone!

Artificial Insemination, Part 3:
Completing the Project

O ver the last eight months we have been involved in our first attempt to use artificial insemination (AI) on some of our ewes. Previously I have attempted to outline what we have experienced in the overall process. I am now able to write the concluding chapter of this experience. Hopefully, I can also come to some conclusions about the value of the process to the flock and to us, as total novices.

I last wrote just after we had performed the laparoscopic surgery to inseminate thirteen of our Corriedale ewes. Two weeks after the AI surgery, each of the ewes was placed in one of our regular breeding groups. It was an intentional back-up system, designed to assure, if the AI did not take, that at least a ram would still make sure that the ewe would be pregnant this spring. Since we use marking harnesses on all of our rams, this normal breeding period was a way to judge how well the AI may have gone, hopefully within one heat cycle from the AI date. The rams did not fail to enlighten us about our successes and failures.

We had used semen collected from three different Australian Corriedale rams. During the insemination process, we experienced difficulty with the semen straws from one of the rams. Our AI technician, Randy, was not too optimistic that this set of semen would end up being viable. Our rams confirmed Randy's assessment, by clearly marking every one of the ewes from this first AI ram. It was a major disappointment, as we knew less than three weeks after the AI was done that nearly half of our project was already unsuccessful. We were also extremely upset by the lack of quality in the product in which we had invested.

Of the eight remaining ewes, on whom we used the semen from the other two rams, four were never marked by a ram and two were only very faintly marked. We knew when we removed the rams in mid-November that at least 50% of these ewes were successfully AI'ed and we hoped for as high as 75%. We had all our ewes pregnancy tested in January using an ultrasound. The process confirmed that all of our 13 AI ewes were pregnant, either due to the AI or due to a clean-up ram. We were relieved that we had not caused any of our chosen ewes to be open for the up coming lambing, since these ewes were the cream of the flock. Our vet's ultrasound machine is not sophisticated enough to project due dates, so it was not possible to tell from the procedure which of our successful AI percentages was correct. The ewes would have to tell us that story in their own good time.

We knew the due date for any of our AI pregnancies was supposed to be about February 20th, i.e. two weeks prior to the start of our regular lambing. We arranged for our shearing prior to the earlier date, and managed (barely) to squeeze in the shearer during a mild (by Wisconsin's February standards) couple of days. After shearing we separated the 6 remaining possible AI ewes to a private pen in which they would have some peace and quiet for lambing. Once shorn it was also increasingly apparent that the two "questionable AI" ewes were not ready to deliver on February 20th. We had suffered our second, not totally unexpected, disappointment.

Kassia, the sole colored ewe successfully AI'ed did not wait until February 20th. On the evening of the 18th, she delivered a lovely set of twin ewe lambs. Odessa and Odetta were both active and vigorous lambs of good size. At last, a moment of great satisfaction! The next day, Justine followed with a single white ewe lamb, Ondine. Twenty four hours later Iowa delivered our first white ram, Oz. Lucky finished off the process a bit late on the 23rd with a set of twin white ram lambs but unfortunately, one of the twins had been dead at least a week or more. The survivor, Omar, is a gem! Every one of the viable lambs was of good size and excellent vigor. There was no need for shepherd intervention to get any of them delivered, up or going. However, disappointment still seemed to be dogging us. Justine's ewe lamb rapidly developed mobility problems. It became apparent that the lamb had some sort of congenital problem, which became worse as the days passed. Despite examination and aggressive intervention by our vet, we were not able to save her. The final total of healthy active lambs with Australian parentage had settled to four.

It has been two weeks since the first of our AI lambs arrived. In a couple of days we start lambing with the other 90+ naturally bred ewes. The brief respite has given us some time to look at and evaluate the entire process and also to reach some preliminary conclusions. Despite the small number, we are exceedingly pleased with the four lambs. They are all doing very well and growing rapidly. It is almost as if we had experienced the hybrid vigor phenomenon. The long genetic separation of the Australian Corriedales (of probably over 100 years) from our own stock may be achieving something similar. If we had to only get four lambs out of the process, achieving two ram lambs from two of our best ewes (who are very remotely related to each other) is a big positive for us. Should the rams prove to be as good as they so far appear, it will permit us to rapidly spread the new genetic branch for our flock.

We are disappointed in the relatively small number of lambs that resulted from initially trying to breed 13 ewes using AI. This small number is further frustrated by the loss of two of the six lambs born. Knowing what we have learned, I believe that we could have done a better job and could have achieved more success. What would we do to improve?

In our situation, the vast majority of our expenses involved the actual insemination process during one afternoon. The cost of our semen was relatively small. If we can find good semen in this general price range, it would behoove us to purchase more straws, from different rams, than we anticipate using in one year. If we had had a different set of semen straws available last fall as a back-up, those straws could have been used instead of the ones which proved defective. We would not have spent any additional money on the synchronization and insemination process, but would have probably doubled our success rate. It would have been much like having a "back-up" ram available should something happen to a ram before or during breeding. If we did not have to use the extra semen straws from the different ram, they would still be available for use in a future year. In such a successful scenario, no investment would be wasted over the run of two years assuming storage facilities are available for any straws not used in the first year.

Had we begun our planning process sooner, we might have been able to make contact with other breeders who had used semen from these same rams. Getting their feedback about their results would have helped. It may have given us some insight into the appearance and performance of the resulting lambs. We might also have been a bit more cautious of reliability of the semen's supplier. Based upon the poor viability of some of the straws and the supplier's unwillingness to acknowledge the problem, we will never deal with that individual in the future.

In the future, we might choose to synchronize more ewes than we actually plan to inseminate. This change would allow us to breed a different ewe should something happen to one of the chosen ewes prior to breeding or should the AI technician not find any follicles in the uterus of a particular ewe. The cost of synchronizing a few extra ewes would not be that great.

Having now gone through the process from start to finish, we should be much more efficient with the same process if we try it again in the future. That efficiency should save some money (e.g. less time the vet needs to be at the farm during insemination). It would also lessen the stress on the ewes, by shortening the time spent on the individual AI, especially pre- and post-operative. Being familiar with the process would allow us to make sure that the materials needed to synchronize the ewes would be available well ahead of time, thereby maximizing the ewe's receptivity for breeding.

One of the down sides to our AI breeding scheme is that our AI ewes were bred two weeks prior to the rest of the flock. As a result our lambing window is now extended (even though there is a break of about 10 days between the first and subsequent group). It means that we are starting to lamb in potentially colder, less hospitable weather (at least for Wisconsin). On the positive side it does permit us to keep a much more watchful eye on the AI ewes prior to and during their lambing.

I hope what I have rambled on about over the course of the last three essays may be of some help to other shepherds, who like us, have never had the opportunity or courage to try breeding their flock using artificial insemination. As this spring progresses we will, most likely, refine some of our ideas. If we have access to good, viable semen from superior rams, I think that we would and will do this again, in the belief that we will have much better success in the future. For now we are just going to savor watching Odetta, Odessa, Oz and Omar doing the wonderful things that lambs always do. We hope all your ewes and lambs are well!

An Ode to a Greeter Sheep

The Icelandic breed of sheep have a genetic line of "Leader Sheep". We have "Greeter Sheep". It is probably not a genetic trait or, if there is one, it is not as obvious as the characteristic Leader Sheep. I have been thinking along these lines recently, because we lost Fuzzy Bear this spring. She was our original Greeter.

What is a Greeter Sheep? One of the prerequisites for such a creature is having a flock that is habituated to a fair amount of day-to-day human contact of the peaceful sort. Within such a flock, at least some of the sheep must be willing to voluntarily approach the shepherd and other humans accompanying the shepherd. Our friends with large flocks or mobs of sheep, such as found in the larger open areas of this country or in Australia or New Zealand for example, will perhaps liken this friendliness to pets. I would submit however that such approachability is based on how the sheep are handled over several generations, regardless of flock size. As our flock has grown, the notion of "pets" has been outgrown by shear numbers. We work our sheep without benefit of a herding dog and move them every day to fresh pasture. For this operation to successfully function, we require sheep that are trusting and willing to work with us rather than react to our actions as threatening. As a result, we tend to select for sheep that will function well within such an environment. Even though our flock now numbers over 100 ewes, we still have a significant number of ewes who display enough trust in us to voluntarily approach us. Their friendliness is rewarded with a scratch under the chin, a back rub and some kind words of praise and adoration.

We have quite a few sheep that, on any given day, will come to visit with us and our B&B guests. It is a good selling point for our business. It also permits us to provide people an up-close discussion of sheep and wool. We hope it is our little contribution toward creating more understanding about the animals and the fiber they produce. This effort is especially important today, as fewer and fewer people in this country retain any connection to agriculture in general and the condition of the family farm specifically. Many of the ewes will calmly permit us to show off their wool underneath their jackets. Of course we have our own motives. If we can convince a few more people to take up spinning then the market for handspinning wool will continue to prosper.

Despite a large percentage of ewes who will visit with us on any given day, we normally cannot rely on specific sheep to respond to us every time we go out on pasture with them. Friendliness by itself does not make a Greeter Sheep. The Greeter Sheep must be relied upon to always voluntarily come to the shepherd (especially when accompanied by strangers) without us calling or otherwise encouraging them. Regardless of where she is in the pasture, she will head our way when she knows we are coming. Without fail she will always be among the first to arrive for such a visit. Quite often she will vocally announce her arrival. Once she arrives she will make the point of personally greeting and visiting with each person in the group. She is a public relations genius, who seems to be able to manipulate each person who visits her. She could easily run for public office and win handily, but she is much to wise to do so.

Fuzzy Bear was our first official Greeter. She met all of the above standards and carried out the role with flair. It was a role she took upon herself early, as a lamb nine years ago, and did not relinquish until her death this spring. It was not a role for which we trained her, but obviously she was rewarded as she perfected the position. She did not come by the job based upon the behavior of her mother Sugar Bear. Originally Sugar Bear was standoffish and only warmed to human contact when she found that her lambs were getting lots of attention and enjoying it. Over the course of her lifetime, Fuzzy Bear could be counted upon to always appear when folks walked out to the pastures to see and meet the sheep. No matter where she was in the pasture she would make a beeline to the visitors as soon as she spotted us. Once arrived, she would put up with people who did not know how to act around sheep. She would always tolerate having her jacket lifted so that people could witness the effect it had on the quality of her fleece. If she felt neglected, she would quietly saddle up to someone and then lean on them. That was guaranteed to get their attention! She was our public relations star. I am sure that more people remembered her name than ours. As a result, she was also a marketing gold mine. One year Gretchen kept her fleece to spin for herself. The resulting yarn ended up in handwoven scarves and knitted mittens. When people discovered them to be made from Fuzzy Bear's fleece, they would inevitably be sold to her new found friends. Hers was a lovely fleece, but not more so than many others from our flock. But products made from her fleece had a special aura that required that they be purchased as a memory of good times and warm friendship.

Fuzzy's behavior/performance was not the same as that of a bottle lamb (or bummer) who, as a young lamb, tends to bond to the provider of her bottle of milk. Our relationship with bummers tends to be that of a foster parent and provider of food. We have had our share of bottle lambs, some of whom are still with us many years later. They are our devoted friends, but it is a friendship that is not often transferred to strangers. It tends to be more of a family friendship, a devotion to us as "parents" more than shepherds. If any friendliness from our bottle ewes is shared with others, it is not predictable.

Fuzzy was not a leader or a boss within the flock. Her place within the social hierarchy was somewhere in the middle. She was never the first one out of the barn or the first one down the lane to new pasture. She never elicited deference from the other sheep as some of our matriarchs have done. While she was always friendly with us (even when we did not bring her visitors), she was not always easy to deal with. She hated to have her hooves trimmed and would continue to fight you to the very last snip. Shearing was only slightly less detestable for her. She was a devoted mother, who often produced twins or triplets, most of whom were males. Unfortunately, as she aged her capacity to raise triplets on her own diminished, due to lack of a sufficient milk supply. We retired her from breeding a couple of years ago because of that problem. She became a member of the "old ladies club" and was expected only to produce a nice fleece and a pleasant, friendly outlook. It was especially difficult this spring when she slowly began to loose an active appetite and lost more weight than her appetite could account for. The vets could not find anything. Despite her ability to communicate with so many people, she was unable to describe her ailments for us. That ultimately is the true frustration a shepherd or any animal caregiver feels when trying to help a sick or injured animal. Eventually, for her own comfort we decided to have Fuzzy put to sleep. It was one of those painful, difficult decisions that rationally made sense, but emotionally pulled at every heart string.

At present we await the emergence of a possible replacement Greeter Sheep. Jorgina seems to be a likely candidate. She appears to be taking some of the initiatives that Fuzzy had in the past. It is, however, too early to know if it is a role that another sheep will take upon themselves with the gusto and dedication that Fuzzy brought to the task. As with the loss of any friend, we currently suffer an emotional void without Fuzzy Bear in the flock. She will never truly be replaced, no matter how good the new Greeter Sheep.

Fuzzie Bear

Swallows and Sheep

S oon the fall equinox will be upon us. In northeast Wisconsin the signs of the upcoming season are numerous. The soybean fields are turning yellow. The last of the late cuttings of hay are off the fields and in the barns. The small grain harvests are complete and corn is just beginning to be cut for silage. Nearly all of the lambs that we are selling this year have now left the farm. With them have gone a small group of older ewes, making room in the flock for younger replacements. The culling of older ewes is, for us, one of the least pleasant experiences and it is one of the few dark moments in what is otherwise an enjoyable time of year. Their departure is somewhat softened by the knowledge that we have a core group of replacement lambs that are doing well and endearing themselves to us daily.

The days are noticeably shorter and as a result the rams are becoming amorous and smelly. It is also apparent that the ewes are beginning to cycle for breeding. It is amazing to think that all of these activities and physical changes are brought about simply by the gradual shorting of the days. In less than a month we will turn the rams in with the ewes and then we begin the entire wonderful waltz toward lambing. We are especially eager to see how the two ram lambs, Oz and Omar, will perform. They are the two rams that resulted from last fall's experience with artificial insemination. So far they have met all of our eager expectations. The arrival of their first lambs next spring will be especially exciting.

The fall season for us is thus a time for departures and arrivals, endings and beginnings. Of the various natural events that occur at this time, the departure of "our" barn swallows is perhaps the most meaningful to us, at least symbolically. I am reminded of this now, because the last brood of young swallows just left their nest high in the rafters of the hay mow. This year's hatching of the last brood is one of the latest that I can remember. They are quite possibly the third set of young this adult pair of swallows has produced this year.

That I write about the barn swallow (*Hirundo rustica*) for a publication devoted to sheep, goats and fiber is really not that disconnected. I look upon the barn swallows as our farm supervisors. They are perhaps more a part of our flock and its operation than any other animal on the farm, save the sheep themselves.

We are lucky enough each spring to welcome back three different species of swallows that will nest in and around the barns. The tree swallows are usually the first to arrive in early April. They will nest in boxes set out on the pasture fence posts. When they have brought off their brood in early summer they will depart soon after. The cliff swallows follow the tree swallows' arrival by about a week. As soon as mud is available they set about repairing their mud nests from last year or building new ones, all located under the eaves of the barns. They will remain most of the summer, entertaining us with their curious squeaky vocalizations to their young.

Nearly coincidental to the cliff swallow arrival is that of the barn swallows. Usually they begin arriving just as we have finished lambing, at a time when we have just begun to relax a bit.

Their arrival will be announced one bright spring morning, usually while we are doing morning chores. It tends to be one or two individuals dramatically swooping through the open barn doors and flashing through the barn, jabbering away with their excited twitters. It is difficult to know whether their animated chatter is a commentary on the condition of their old summer home or a statement of joy that they have finally returned home. It is also as if they are making their first assessment of the new lamb crop and at the same time passing judgment upon us and our care of "their" summer home. Over the next week or so more and more return. They too will begin to set up house keeping throughout the barn. By the end of this summer we had a total of 29 nests in the lower "sheep" level of the barn, 9 more in the new addition and, at least 6 near the very peak of the roof up in the hay mow. How the later group survives the heat of midsummer in that spot is a source of total amazement to me.

The barn swallows will accompany every phase of our spring and summer with the sheep. Upon their arrival and during the early stages of nest building and/or repair, they seem to have a special interest in the goings on of the sheep. Often they will land on either the old pegs that once held the draft horse harnesses, or sit on the tops of the panels separating the various groups of ewes and lambs. As they sit and watch, they converse away in merry fashion, seemingly commenting about the various young lambs as they develop. In turn, their serenade will be a constant source of curiosity for the lambs, who will watch the swallows for hours.

By the time that the pastures have greened up enough to begin grazing in early May, the swallows will have set up serious house keeping, with many of the nests already filled with eggs that the female will have started to incubate. The departure of the flock from the barn in early morning generates an aerial exhibition by the swallows complete with vocal commentary. Almost as if they are wishing the sheep and lambs a good day on pasture, they accompany them out the door and on to pasture. At dusk the procedure is reversed, until everyone, sheep and swallow have settled into the barn for the night.

As the summer progresses and the swallow chicks begin to hatch, we humans in and around the barns are perceived to be dramatically different by the swallows than are the sheep. We seemingly are a threat. As a result, we are the target of aerial attacks, rapid strafing assaults from all directions which are broken off at eye level at the last possible instance, so close at times one can feel the air from their wings and hear the click of their bills as they fly by, excitedly cursing our very existence. Even after years of these assaults, I still flinch as one of the irate birds swoops straight at me. If they are keeping score in this game of "chicken", their performance rates the perfect 10. Conversely, the departure and arrival of the sheep and their continued presence in the barn at night is happily accepted by the swallows.

Once hay is ready for first cutting, the swallow chicks are big enough to demand a good hefty feeding. How four or five clamoring chicks can all stay put in their teacup sized nest is amazing. For once the adults and I are on good terms (away from the barn) as I cut, rake and bale hay. They are my constant escorts as I slowly circle the big field in the tractor, swooping around to grab the numerous bugs that the tractor and hay equipment have put to flight. Besides having the enjoyment of watching their aerial acrobatics, it is just plain pleasant having some company in the long hours spent alone in the field.

As the young take flight, they too will join in the air show centering on haying. The young will often perch on one of the perimeter fences along the edge of the hay field or on the back of empty hay wagons parked in the field, where they will beg and clamor for food each time an adult flies by. At this time, the morning departure of the sheep from barn to pasture becomes a grand parade of sheep and lambs on the ground with swallow escort in the air. The swallows enjoy the lush bug breakfast stirred up by the sheep upon their arrival in the pasture. I have no way of documenting it, but I feel confident that the swallows return the favor by significantly reducing the mosquito and fly level that would otherwise plague the flock.

The swallows' reproductive cycle will often be repeated twice and occasionally three times during the summer. Often the occasion for cutting second crop hay means that I am again not alone on the big hay field. But by late summer, some of the families and their offspring begin to depart, at much the same time as some of the lambs begin to head for market or new homes. Slowly the barn begins to be less full in the evenings, both in terms of swallows and sheep. This year by the end of August all the nesters and young from the lower areas of the barn had departed. In early September only a couple of nests remained active in the hay mow. Is it coincidence that the level of flies around the barn at this time begins to increase? I think not. Our swallow armada, up until now had done a wonderful job of fly control.

As I write, the last nest in the mow has fledged. The adults are all in a tizzy when I climb up to get some straw bales for bedding, while the young try out their wings for the first time above the summer's accumulation of hay bales. Just as our last large group of lambs left the farm, so too have the barn swallows. The interior of the barn suddenly is deadly silent.

It is my understanding that the barn swallows in North America will head for Central and South America to spend our winter. I fantasize that the group that spends their summer with us will soon be arriving at a sheep ranch, perhaps in Uruguay or Argentina. There they will take up residence and pass judgment on the Corriedales they see there. Perhaps their twittered conversations at times center on discussion of whether those sheep look better than ours, have nicer fleeces or cuter lambs. It is of course just my dreaming. Nevertheless they are truly a welcome guest at the farm each year. I believe that somewhere in Europe, the same swallow, my little friend, *Hirundo rustica*, goes by a common name that often translates into Welcome Swallow. I have fond memories of their happy presence in a dairy barn in Denmark. Even if the name "Welcome Swallow" is merely the product of my own imagination, it is still appropriate. With the departure of the last family this month, we already look forward to their return next spring. It will be an event nearly as joyful as lambing, and lambing would not be complete without the arrival of our swallow friends.

Trolls and Phobias

Winter has arrived at the farm. While the first good, heavy snow is yet to fall, the cold temperatures and brisk winds off the lake confirm the season. Because it is winter both our own routine and that of the flock is altered significantly from that of the other three seasons. For us early winter is the one time of year when we can relax and do many things only dreamed about during the busier times of the year. Breeding is over for the ewes, shearing is still two months away and lambing is still a little farther down the road.

The colder weather of the season, of necessity, changes the routine for our flock. Pastures have been well grazed and are now dormant until spring. The prospect of a good meal of fresh grass grazed from the midst of a field is now only a dream. With the cold weather, the path that slopes out and down from our old barn, which is built into a hillside, takes on a different character. Traversing the path becomes an adventure when it snows or ices over. It becomes impossible to drive any type of equipment up its slope under these conditions. It is a challenge for us two legged animals to walk up or down. It is not even a pleasure for the sheep to navigate, especially as they get close to lambing. As a result of all of these conditions, life for the sheep, in the dead of winter, centers on the barn and the area immediately around it. It is shelter from the worst of the weather and the place in which fresh bales of hay miraculously appear twice a day. Were they given the chance to voice their opinions of the conditions, the flock would probably, as a whole, have to admit that they have it pretty good at the present time.

However, living a life of leisure, in a comfortable environment, does not mean that life is not without its trying moments. We all have our phobias of one sort or another. Why should sheep be any different? At the moment the flock is going through its annual sunrise phobia. The exposed portion of the sheep quarters faces east. With the maple grove east of the barn now devoid of leaves, the sun rising from the southeast on a clear morning floods the barn with bright, warm light at its lowest possible angle. From our perspective, it is a wonderful, cozy light that makes a beautiful start for the day, especially if the trees and pastures are aglitter with frost. There is unfortunately a dark, sinister force at work here. Great shadows suddenly appear on the west wall. As a ewe moves toward a feeder, the shadow follows and descends upon her and the feeder, always at a greater speed than the ewe's. Its height is always larger than the ewe's and is all the more intimidating. Even if the ewe is unaware of the great danger she is in, there is sure to be another ewe who takes note of the threat and sets off the silent alarm. Once the alarm is triggered, security can only be found in the south end of the barn where the evil, two dimensional creatures are unable to lurk. God help the poor unsuspecting shepherd, dutifully filling feeders at this moment, who is totally unconcerned about his own even larger shadow and who is suddenly trapped in the midst of the ovine rush (having been without benefit of the unspoken alarm call that has passed rapidly among the sheep)!

Sheep have excellent memories. This quality is obvious as the winter progresses. The ewes know where the shadows live and lurk along the west wall. So even the slightest movement along that wall (even on the cloudiest and darkest mornings), is confirmation that the creatures are still present but merely being more secretive. If one remembers where the creatures come from, it is also much easier to notice that at evening chores, when the shepherds turn on the lights in the barn, that the shadow creatures are still there. They have just assumed a smaller size and different identity. It is amazing how such sinister creatures can transform themselves into little beasts that look surprisingly similar to mice! In that disguise they are able to hastily make their retreat into holes in the same wall. Nevertheless, the ewes are now on alert and are not fooled by the simple mouse costume. They know better and therefore beat a rapid retreat to the south end of the barn.

I imagine that when the ewes are finally safe and secure they inevitably discuss the powerful forces that threaten their daily existence. Through these discussions, those tales are no doubt passed down through the generations. How else can one hope to explain the fact that every spring the vast majority of the flock remembers where the troll gate is located?

Trolls hibernate in these cold climates, but once the weather has sufficiently warmed to start the grasses growing, the trolls emerge from their winter snooze. Being the nasty lot that trolls are, they like nothing better than harassing the sheep just as severely as their cousins the shadow creatures. They know that the sheep must pass along side the old stone fence line nearly everyday, as they go out to pasture in the morning and return at night. And so, the trolls have set up a gate along side the stone wall to stop the sheep in their daily travels. Once the gate is closed, the sheep will not and cannot pass it. It should also be pointed out that the trolls are able to make these gates invisible to the human eye of the shepherds. However a sharp-eyed shepherd can learn to identify the signs of these gates even though invisible to shepherd eyes. Usually the most reliable indicator that a troll has just slammed the gate shut is the fear it instills in the smaller creatures who live in the stone fence. As the gate slams shut, it frightens a mouse or chipmunk and, in their fright, the little creature will run across the path in front of the sheep. The flock knows and remembers these signs. They remember that a chipmunk surging out of the stone fence means a troll is near and is about to close the gate. The flock will remember from one year to the next where these poor chipmunks appear and as a result will not pass that spot for days unless the shepherd is brave enough to first work his way to the front of the flock and is able to frighten the troll into hiding and in the process push the gate open.

In all probability, the trolls will appear elsewhere around the pasture, likely in retaliation for the perceived mistreatment by the shepherd. There are days in which the troll will hide up in a tree and out of malicious spite try to frighten the sheep. They will achieve this by placing a large stick or branch across the path traveled by the sheep to the pasture. When the sheep come across the stick, knowing that it was not there the last time they passed, they will refuse to travel further until someone (a.k.a. the shepherd) removes the stick and checks to make sure the beast is gone. Given enough space to the side of the stick, a few brave sheep may try to pass it if they can give it wide enough berth to be able to rush past as fast as possible. Usually the evil creature has carefully calculated the placement of the branch to cause the greatest difficulty in passing. The result is a few of the mob of sheep bouncing into and off of the electric fence at their side and then back into the middle of the path, thereby causing further anxiety and haste within the remaining flock.

As the sheep head out to pasture each day, I suspect that these trollish creatures are also responsible for occasionally grabbing at one of the coats the sheep wear. Every once in a while they manage to cause the coat to rip or come loose so that it will drag on the ground beside or underneath the sheep. Panic can understandably escalate, if the jacket does not come off, but rather follows the sheep around. It would seem that the creature is riding right along. Why else would it be such a fearsome sight? If the jacket comes off, in the pasture, the troll obviously sits atop it until the hapless shepherd recovers it. Until that time, the sheep know very well not to go anywhere near the crumpled pile of fabric, even if there is good grazing to be had just a few feet away.

Strangely, most of these dastardly tricks are at the expense of the ewes and their lambs. The trolls and their kindred creatures seem to give the rams a wide berth. The shadow creatures never seem to occupy the rams' quarters in the barn. Rarely do they even leave obstacles in the rams' way as they head out to pasture. Perhaps it is due to the larger size of the rams or their often aggressive manners. The ewes may also suggest that the rams are also just a bit thick headed! I can only speculate, but even trolls would probably not be dumb enough to get in between two rams as they are seriously butting heads. On the other hand, perhaps it is due to the excessive odor that seems to follow the rams wherever they go. It is, after all, a smell that seemingly only a ewe can love!

As lambing time approaches, we spend more and more time each day with the sheep in and around the barn. To a significant degree this seems to deter the shadowy creatures from causing too much havoc. It is hard to say whether this is due to our presence in the barn or just because the ewes become preoccupied with lambing and their new born lambs. In any case, it is probably time for me to head for the barn to look in on the flock. It is a good thing it is cloudy now. I need not be too concerned with that shadow which seems to have been following me to the barn recently....

A Normal Lambing

I had hoped to begin writing this piece a couple of weeks ago. It was at the very beginning of lambing in March 2004 and there should have been time in between the scattered deliveries to get some words written down on paper. After all, this was just going to be another "normal lambing". But here it is, now just over three weeks later, and I've not finished the first paragraph. The events of these last three weeks have made me wonder whether there is such a beast as a "normal lambing."

Needless to say, each shepherd has created their own unique situation when it comes to lambing. The size of the pregnant ewe flock plays a significant role. The duration of the lambing will determine the amount of stress and fatigue experienced by shepherd and flock alike. In a similar fashion, how the lambing facility is set up and how "friendly" it is for man and sheep can be crucial. Most of us probably have limited, if any, extra labor to assist us during lambing. The season of the year that has been chosen to lamb is another significant variable, as is the location. Lambing near the Artic Circle produces different challenges than a warm, dry temperate location. From my perspective, I am only able to relate to lambing in relation to what occurs in our little 80 acre corner of the world. For us a "normal lambing" has been determined by the following: 1) The time of year: the month of March into early April, a period transitioning from snows and single digit Fahrenheit temperatures into early spring and days warm enough to make the grass grow.   2) The number of pregnant ewes and ewe lambs: usually from 75 to 90. 3) The time that transpires from the first lambing to the last delivery: six to seven weeks.

Over the years we have chosen to breed our ewes from early October to mid November, usually allowing for about three heat cycles for the ewes. Our breeding dates would, hopefully, allow lambing to avoid the coldest, least hospitable weeks of winter. Our hope is that with lambing scheduled to begin around the first of March, that the days would be getting warmer and the weather becoming more pleasant for both us and the sheep. Then by the end of lambing, 50 days or so later, the grass would be starting to green up so that within a week or two the flock would be out on pasture for serious grazing and not dependent upon winter's store of hay.

With those targeted lambing dates, we are also able to schedule our shearing for about 2 weeks prior to the first's lambs arrival. Over the years mid February gives us at least a brief warmer spell, which permits us to shear the flock and then keep the ewes in the barn to stay warm until lambing begins. It seems that as part of our "normal lambing" we usually find ourselves spending Valentine's Day in the barn shearing some of the flock. The two weeks between shearing and lambing also permits us to get most of our fleeces aired, sorted, re-skirted, weighed and priced. With luck we are ready to put them up for sale shortly after we complete lambing.

We make no effort to synchronize our ewes' heat cycles, either through chemical means or by separating the ewes far from the rams for a long period or by using a teaser ram. This lack of synchronization means that the ewes will presumably come into heat on a random basis over the 17 day heat cycle and hopefully get bred within that time. Should they miss the first cycle, a smaller number of ewes will get bred during the second 17 days, with a few stragglers getting bred (or re-bred) during the last of the three 17 day periods they are with the rams. The result usually is that the vast majority of our ewes will lamb spread out over the first 17 days of lambing, with a smaller group following in the final 34 days. Traditionally, the late group tends to consist more of ewes who were bred as lambs, by a ram lamb. Young age and experience (or lack of it) means that they tend to take a bit more time to get bred successfully. We can, therefore, usually expect to have to deal with inexperienced, first time mothers after the initial crush of deliveries have passed. Hopefully, it also means that we are able to devote a bit more attention to these first time moms than if they delivered in the early stages of lambing.

Having lambing spread out over these 40 to 50 days is logistically helpful for us. Since the chosen time for lambing is still winter (at least climate/temperature wise), we must contend with the birth of our lambs indoors in the barn. We have sufficient space that we are able to set up as many as 15 jugs for the ewes and their newborn lambs. Spreading out the due dates for 90 ewes, usually allows us to give each ewe three or four days in a jug with her new lambs prior to going into a mixing pen with other ewes with lambs of similar age. Having those three to four days in the jug permits us to spot potential problems which otherwise might get missed in a larger pen of ewes and lambs. It also gives us time to trim the ewe's hooves and fit her into a clean jacket, thereby allowing the lambs to become accustomed to their mom in a coat before they also have to contend with a mass of other ewes and lambs.

There is, however, a down side to a lambing spread out over 40 to 50 days. The lambs that are born late have a distinct size and age disadvantage when competing with the older lambs in a creep. The logistics of getting a two week old lamb to walk out to our more distant pastures each morning is daunting for the lambs and their mothers (not to mention the shepherd). It is actually amazing how well they do with the situation. However, the single greatest problem with a lambing period this long has not so much to do with the ewes and lambs as it does with the shepherds. When there are just the two of us to handle all aspects of lambing, the fatigue factor becomes significant, as the days run on. The older one gets, the more difficult it seems to be able to go for over a month without a single night in which we can sleep the entire night uninterrupted by midnight "bed checks" or early morning deliveries.

Because of all of the above conditions, our lambing routine has been an evolutionary process. Each year we learn a little more from our mistakes, and make note of what we did that worked well or what failed miserably. Each lambing is similar to the previous, but hopefully is perhaps run a little better than the last. Unfortunately, one cannot completely plan for the weather (e.g. prolonged periods of extreme unseasonable cold), or how many ewes and/or lambs may develop health problems. Despite the vagaries, we have become better and more efficient midwives than we used to be with the lambing operation. We were made more aware of this when we lambed a year ago. In 2002-2003 we intentionally "messed up" our normal routine by artificially breeding some of the ewes. We did the artificial insemination (AI) prior to the date we normally would place the rams in with the ewes. Our intent was to cover any unsuccessful AI with a natural breeding at the usual time. It proved to be a good policy since we had only about a 50% success rate with the AI procedure. Those ewes that did not get bred using AI were still successfully bred by our rams at the normal time. The plan worked well, but it extended the time that we were lambing by almost three weeks. There was a short break in lambing after the first week, but by then our psychological clocks were set for lambing. The extra three weeks that we added to our "normal lambing" schedule ultimately took its toll on us. When that year's lambing was over, we were tired, without a doubt.

All this brings us to the current lambing season. With the extended lambing of the previous year still fresh in our minds last fall, we made a couple of management decisions that would hopefully make life a little easier for us and the ewes come lambing this spring. In the hopes of tightening up the time we spent lambing, we cut a week off the time we left the ewes with the rams in the fall. Obviously we therefore knew that we would be done with lambing sooner. We also calculated that it would mean that fewer of our ewe lambs would successfully get bred their first year. Thus, we also figured that we would have fewer pregnant ewes to deal with during lambing. Our goal was to simplify "normal lambing".

Even though we are not yet quite done with lambing as I write, I can already report that we attained our goals of shortening the lambing period and reducing the number of bred ewe lambs. However, it has proven to be far from a "normal lambing". When we finished with breeding, we had an indication that things would be different this spring. Despite the fact that we had done nothing different with the ewes and rams prior to breeding, the initial week of breeding was a madhouse. We almost were beginning to wonder if someone had slipped the ewes some heat synchronization drugs. Over half of the ewes were well marked by the rams within the first five days of breeding. While the breeding pace eventually slowed a bit, it was still becoming obvious after the first heat cycle that most of the adult ewes were successfully bred during the first week of breeding. When due dates arrived in early March, we were, at least, ready for the onslaught. As it was, it started two days early. Nora, one of the ewes bred to Oz (one of the Australian AI offspring) started things off fine on the first of March. And then we waited…and waited. Lambing finally began in earnest: three ewes on the 4th of March and 2 ewes on the 5th. At that pace however it seemed "normal"; there was no big rush to lamb half the flock in the first couple of days. By the 6th the flood gates opened. At the end of three weeks, when the dust had settled, all but three of the ewes had lambed. Somehow, early in the process, the notion of leaving each ewe and her lambs in a jug for three or four days just evaporated. The jugs should have been fitted with revolving doors! Surprisingly, the ewes and lambs coped well. We had definitely tightened up the lambing time frame.

It was also apparent that we would not have as large a percentage of the yearling ewes bred as in the past (which was as we had expected). We were, therefore, faced with being done with lambing three weeks sooner than we were the previous year. The total number of live lambs born will not be that much different than the previous year. We have been fortunate with survival rates, having lost only one lamb this spring out of more than 120. Having the ewes lamb so close to one another on the calendar has meant that their diets have been more uniform and the lambs healthier and more rigorous at birth. The ewes have also had fewer health problems with which to deal.

Try as we might, we could not influence the weather. Despite what has turned out to be a very successful lambing, the weather has been fickle. It is only spring on the calendar. Snow still lies heavily on the ground and temperatures have been cold, highlighted by extreme dampness from the east winds blowing off Lake Michigan. While we are going to be done lambing early, the warmth of spring is staggering in this direction at a much slower pace. The ewes and lambs will be ready for life outside the barn earlier than in the past. It just remains for Mother Nature to pick up the pace.

Being a young lamb can be serious business

About 120 lambs later, it has been a very good lambing, albeit very hectic. It will not be as "normal" a lambing as all the rest, but it is a definite improvement. Someday, perhaps, we will look back upon this year as being the first of many "normal lambings".

Why Do You Raise Sheep?

Sometimes I wonder about the various reasons why many of us raise sheep, goats or other fiber animals, especially animals of the colored varieties. I am sure that the reasons are diverse and numerous. The longer one cares for a flock of sheep, the underlying reasons for one's attachment to them becomes more ingrained, perhaps to the point where one rarely thinks consciously about the causes. Occasionally there will be events that jolt my mind into thinking more actively about these basic sources for inspiration in raising our flock. Most recently, the event that caused this stream of thought had to do with my "fussing".

What is it I have to fuss about? It has almost become an annual "spring fuss", related to customer relations. Over the last few years, the majority of our fleece sales have been the result of our concerted efforts online through the internet. It has proven to be an excellent method for us to market our annual clip. It is a very intense and busy time for us, however, especially since it always dovetails with the end of lambing, a period in which time is already a precious commodity. Our customers are, as a group, very wonderful, very cooperative and very understanding. We have, as a result, become good friends with many without ever having met them. All of which makes it a rewarding personal experience (and therefore adds to the list of good reasons for raising the sheep to begin with). However, it seems inevitable that there is at least one buyer who goes against the otherwise pleasant and enjoyable flow. Getting them to pay for their order and to provide the necessary shipping directions becomes an unpleasant ordeal. If they eventually come through with a payment, we may still decide not to keep them on our mailing list for the next shearing due to the aggravation and trouble they have blessed upon us. Occasionally, after placing their order, for no apparent reason, and before providing us with payment, they fail to respond to any subsequent communication. This behavior then requires us to put the fleece(s) back up for sale. All of which causes me to fuss about what I perceive to be their strange and discourteous behavior.

I believe that this type of customer behavior is largely attributable to the urban life style and mind-set that our society seems to foster more and more. The gulf between an urban and a rural/agrarian life sees to grow a little with each year. It is a gulf that I am sad to see growing larger. The concept of making one's living off the land is decidedly different than being accustomed to a regular pay check. The customers about whom I fuss, probably have no sense that paying for their fleece order is at all similar to their receiving a regular pay check. In addition, there seems to be no pride in their own honesty or dependability. Thank goodness this behavior has not become the norm; selling fleeces would otherwise become an onerous task.

In a round about fashion, I am now back to my original quandary as to why each of us raises our special flock. In our personal situation, it is a multi-faceted answer. (I will only touch on a few of the reasons now, perhaps saving the rest for another time or for someone else's discussion.) We have a personal love for our animals. We also have a deep respect for them, their lambs, and the lovely fleeces they produce each year. The excitement of seeing new lambs born each spring is still very real, especially knowing that they may have the "ultimate" beautiful fleece. That they can be so dependent and trusting of us is humbling. Our ultimate reward from this endeavor is being able to support ourselves in a rural agrarian environment using the flock as a major underpinning. Having had this experience, I cannot image a return to an urban life like we once knew.

It is a blessing that so many of our fleece customers still have respect for this rural farm environment. On occasion, I need a gentle reminder of some of the standards that support this way of life. It is gratifying that one can have a relationship with one's neighbor that is largely built on trust. For us, it is still possible, for example, to purchase a piece of farm equipment (perhaps a baler or even a new tractor), based only upon a handshake. No advanced down payment is asked for or expected, since Len or Dean know that when the implement arrives and is delivered you can be trusted to pay for it. It is the type of trust that accompanies doing business with the same folks for 10 years, 20 years or a lifetime. It is also one of the complex and subtle reasons that we feel good about raising sheep the way we do.

I also believe that there are very personal gratifications from the raising of our sheep that transcends both the love of a rural lifestyle and the personal love of individual members of our flock. Being able to share the joys of the sheep with others and to see them glow with excitement and satisfaction over these discoveries is a great reward. It is wonderful to watch a person who has newly discovered the joys of working with fiber, learning to spin it. This enthusiasm can apply to both the very young and those of us not quite so young. I was recently reminded of this by the death of a good friend. Galen only took up spinning when he was in his 60's.

But once he discovered its joys, it became a passion and a creative outlet for the self-styled "crotchety" Norwegian. He would not only spend hours behind his wheel honing his skills, but would relish the opportunity to have a quiet visit with the sheep that supplied the wool he spun. Some of the sheep and lambs were especially and uniquely drawn to him; they seemed to share a mutual respect. It was rare that Molly, one of our ewes, would not make contact with Galen when he would arrive in the barn. They seemed to willingly carry on a conversation in their own private, non-verbal language, a conversation that seemingly would have lasted for hours had the opportunity presented itself. Being able to be a part of Galen's discovery of spinning and wool is a reward that needs no monetary value. The fact that we have lost Galen this spring has left a void that will be hard to fill. In their own way, Molly and her ovine friends will miss him just as deeply.

Finally, on this short and incomplete list of reasons why we raise our sheep are the numerous personal rewards that the sheep have given us and which they continue to provide. Much like the personal friendship provided by a human friend, individual members of the flock are able to provide us with a personal devotion and trust that cannot be measured financially. They are able to remind us of our own weaknesses and our inability to understand their complex personalities, while at the same time providing as warm and as close a relationship as is possible. That we are able to still achieve such a relationship with many members of the flock, when it numbers over 100 adults is especially gratifying. This type of relationship was epitomized by Emmy Ewe. That we lost her, just as we lost Galen, this spring is somehow not just coincidence. Emmy had a singular way of reminding us of our own imperfections and the fact that we knew much less about sheep than we cared to admit. Emmy Ewe was one of the earlier lambs born on our farm and as such she was born when our sheep learning curve was still extremely steep. In actuality, she began teaching us prior to her birth. It was because of her unscheduled arrival that we learned that four month old Corriedale ram lambs left with the adult ewes were able to successfully breed more than one of them. As such, Emmy was born three months before the scheduled lambing dates. We were never sure who her father was, but most likely it was Darryl, or his brother Darryl (that is not a typo!). Emmy had a striking resemblance to both of them. Emmy Ewe had a good and devoted mother, Beatrice. Yet even though this was Beatrice's second lambing we were concerned because we never saw Emmy nursing. Beatrice seemed to have very little milk, so we tried supplementing Emmy's diet with a bottle of milk (which she refused). As last resort we tube fed her for a number of days. We stopped that procedure abruptly when, after weighing Emmy, we realized that since birth she had gained considerably more weight than the combined total of all her tube feedings. It seemed that Emmy Ewe and Beatrice preferred to dine alone. It was a truly humbling experience for novice shepherds! Either because of this treatment or despite it, Emmy Ewe became a lasting close friend for both of us. She was a good mother, who successfully raised many lambs until her milk supply finally started to dwindle. By then, she had endeared herself well enough that she earned a retirement slot with the flock. By last fall, it was apparent that an arthritic shoulder was causing her serious discomfort. Going on long walks to the pastures this spring would be problematical. Then quite suddenly, in mid-winter, she went blind in one eye. Six weeks later the second eye began to noticeably fail. Despite her ailments, she remained a good friend. She loved her extra handful of grain and would loudly announce each evening that she wanted her hand-out. Despite her handicaps, she would usually work her way through the large crowd of younger ewes to claim her reward. By late spring, it was apparent that the sight in her other eye was failing fast enough that she could no longer see the feeders well enough to command a prime location, yet somehow, as if using radar, she managed to work her way to the hay and get a good meal. It was a difficult decision to have her put down this spring, but the progression of her ailment eventually dictated that it was the kindest thing to do. She no longer could comfortably manage the walk out to pasture each morning. Her inability to see the electric fences became a liability. What we learned from Emmy Ewe will never be paid for by all the lambs that she raised, nor will all the fleeces she provided ever cover the cost of her friendship; it was much too valuable.

So, fuss as I may about an occasional discourteous customer and their lack of commitment to a single fleece, the fussing is truly an aberration in the grand scheme of things as far as why we raise sheep. The rewards from being part of the community, from making friendships and sharing dreams with both humans and sheep, are satisfaction in themselves. It may not pay all the bills, but it serves to warm the heart and more than justifies our devotion to the flock. Enough fussing!

Where Did Summer Go?

Early September is here and but it seems as if summer has yet to arrive in northeast Wisconsin. Perhaps it would be more accurate to say that we have just had summer bypass us all together. In many ways, we have merely leapfrogged from an over-extended spring into an eagerly early fall. It seems such a long time since we experienced what was once considered a normal year, characterized by normal seasons. Without digressing into a discussion of who is responsible for these ongoing, atypical seasons, one has to admit that our weather patterns in the upper Midwest have been dramatically altered (as they seem to have been for much of the planet).

The unusual weather patterns have made sheep farming a continuing saga of new experiences and challenges. Just when you think that you finally have the business figured out, nature manages to throw still another dramatic variable at you.

Early spring this year seemed to begin as one would hope and expect. We received a goodly amount of snow and experienced cold temperatures. It was, however, nothing to really disrupt lambing. Lambing proceeded well for the flock. We experienced few problems with the ewes and their new born lambs, which is always a blessing. It was a lambing, however, which produced a larger than normal percentage of ram lambs. When lambing was over about 65% of our new lambs were male (as compared to the statistical norm of 50%). While such aberrations are occasionally expected, it became apparent that it was a phenomenon for many flocks in the area, not to mention many bovine dairy herds. Bull calves were just as common as ram lambs. We had hoped for at least an even distribution of ewe lambs versus ram lambs, since we knew ahead of time that we were going to be keeping a larger than usual number of replacement ewe lambs. Our need for additional ewe lambs was due to a higher than normal turnover in some of our older ewes this last winter and spring. The net result has been that we have retained just about every good looking ewe lamb we produced, colored or white. It was not going to be a good year to offer nice, young ewes for sale as breeding stock. Conversely, it was a good year for selling market lambs, especially since we had excessive numbers of wethers. At least, the market price for lamb has held up well throughout the summer and into the fall.

It was also a spring in which we looked forward to the first home-grown generation of lambs produced by the rams and ewes that resulted from our artificial insemination project from the previous year. We were able to use the two ram lambs that resulted from the AI project. They both did well, especially since (with two exceptions) they were only permitted to breed last year's ewe lambs. The two rams, Omar and Oz, are both white and have strikingly different fleeces. Both of their fleeces are quite nice and have improved upon the quality <u>and</u> quantity of their dams' fleeces (which was the desired goal). It was an exciting time to see each of their offspring arrive and then develop over the next few months. Perhaps, the only downside is that since the two rams were white, all of their offspring were also white, including those whose dams were colored. We had expected that genetic result, but it was nice to dream that there might have been some recessive colored genes

140

floating around in the two boys' Australian ancestry. It will be another year and another breeding, but at least the white lambs they produced with the colored ewes will be carrying enough colored genes that, if we breed them to a colored ram this fall as we plan, we should have at least an even chance for colored lambs next spring. We also had two colored ewe lambs, Odette and Odessa, produced using AI last year (using semen from a colored Australian Corriedale ram). Of the two, one of the two was successfully bred by Omar. Even though the resulting lamb, Queenette, is white, she is genetically still 50% Australian Corriedale, since both of her grand sires are from Australia. Luckily, she seems to have inherited all the good fleeces qualities of both her parents. So far, she is a prodigious wool producing machine: a short legged, chunky, old fashioned Corriedale. She and her mother, Odette, are inseparable and behave more like sisters than mother and daughter. Just like two teenagers, they are constantly into mischief, but they are both true delights.

While watching the lambs develop has been its own reward so far this year, trying to adjust the rest of our schedules to the weather has been truly frustrating. Spring (on the calendar) was unusually cool and wet. The pastures were slow to develop and, as a result, it was later than usual in May before we could begin serious grazing with the sheep. As the spring progressed and transitioned into summer (on the calendar) it did not warm up dramatically nor did it dry out. It just stayed wet! In reality, it was a long, extended spring, well into July. The cool season grasses, however, began to thrive. Once the pasture growth rate got up to speed, it was a challenge for over 250 ewes and lambs to keep up with the grazing. In a normal year, if there was such explosive growth, it is possible to leave a pasture out of the grazing rotation at least once and instead cut and bale it. This tactic usually allows the sheep to stay apace with the lush growth on the other paddocks. The rains continued however. The intervals between storms were rarely more than a few days, short enough that it was extremely difficult to cut, dry and bale any hay. We did not complete the first cutting of hay until mid-July, at least three weeks behind normal schedule. There was so much growth that we did not have enough barn space for all the first crop hay. So much hay and not enough mouths to eat it…! For only the second time in our history, we had more hay than we knew what to do with. For the first time ever, we contracted with a neighbor who does custom large baling. There is a dramatic difference between the traditional small square bales that we are set up to make. While our small bales usually weigh about 40 pounds, the large squares top out in the 800 to 900 pound range. Needless to say, these monsters require specialized balers and handling equipment. Our neighbor was able to put the last 30% of our acreage into 61 large square bales, which we then, in turn, were able to sell. The large bale process was truly amazing, both in terms of speed and efficiency of the operation. Were we a few years younger, it might be valuable for us to remodel and/or rebuild our barn to be able to accommodate the big bales. But, as it stands, we will probably continue to muddle along making large lots of small square bales until we are no longer fit for the task.

We were lucky farmers, being strictly pasture based. Others were not so fortunate. Any farmer in the area needing to plant crops for this year was totally frustrated by the cool, soggy weather. Corn planting was late (if one could even get equipment into the field!). Germination was slow and spotty. The same applied to small grains and soybeans, all of which are still way behind schedule in terms of harvest. The normal warmth (a.k.a. the hot, dog days) of summer never arrived. The only vegetables in our garden that thrived were the peas, who were convinced they had gone to pea heaven where they could bloom and develop lovely, fat pods forever in perpetual spring. Commercially peas are usually finished and harvested here by early July, but this year they were going strong into September.

Interestingly, once the first cutting of hay was finally all baled and off the field, the rain ceased. Had it been the usual hot days of late July and all of August, the pastures would have been cooked. Instead, they stayed green with the continued cool temperatures, but decided to stop growing. When it should have been time to make a second cutting of hay in August, there was very little. Only now, in the second week of September, have we managed to get the second crop hay cut. The sheep have managed their pastures better. By adding some of the acreage that was cut the first time around to the grazing pastures acreage, we are managing to dole out enough good forage for grazing that we should just about get to (and through) breeding time. That situation will be a truly satisfying experience.

Because it had been cool and wet early in the summer and just plain cool for the rest of the summer, it seemed as if we just bypassed summer completely, jumping directly from spring to fall. It is interesting that many of the deciduous trees are responding accordingly. By the first of September, there were signs that the maples and birch trees were already starting to turn color. Some of the migratory birds are on the move early. The Eastern Bluebirds are gathering in their usual large family flocks and following the sheep as we move pastures. The portable tread-in fence posts are ideal perches for them as they harvest bugs from the fields. But soon they are gone. Flights of Flickers and Flycatchers are moving through daily. Our joyous armada of swallows has disappeared. We miss their happy racket in the barn and their very thorough job of fly control. They too seem to sense the early arrival of the fall season.

Hopefully, the hours of day light will still be the trigger for the normal seasonal heat cycles with the sheep, as we certainly do not need them thinking that they are ready for breeding this early. As it is, this fall we will be starting breeding a week earlier than we tend to do. We are not doing it because we fear an early winter, but because we plan to take time off in late fall. We hope to have breeding completed a bit earlier, which will permit Gretchen and me to take some time together away from the farm for the first time in a very, very long while. As yet I cannot imagine what unusual weather nature will throw at us during fall breeding, but perhaps we can get our own seasonal clocks reset with some time off and a different scene.

A Trip "Down Under"

I am totally confused as to how I should begin this narrative. Over the last month, my co-shepherd and wife, Gretchen, and I have had so many new sheep and farm related experiences that it is difficult to keep them all straight. The simultaneous inclusion of a variety of non-farm experiences confuses the images further. The two of us have just spent the entire month of November in New Zealand. Now I am faced with the challenge of trying to share something of this potpourri of experiences without the effort sounding too much like a "How I spent my fall holiday" essay.

The occurrence of the Sixth World Congress on Coloured Sheep in Christchurch, New Zealand was the impetuous for our trip. Besides its exceptional educational value, the trip was, in truth, a true vacation for us. The World Congress on Coloured Sheep occurs every five years. We had dreamed of attending the last two World Congresses, but in each case they occurred in the height of our busiest time, our summer. Each time we heard a report from someone who had been able to attend, we regretted the missed opportunity even further. For us, a Congress that occurred in our late fall or early winter was the perfect scenario. The barn was full of hay. The ewes had completed their breeding period with the rams. The grazing season was over, making sheep care much less complex (i.e. feeding bales of hay in one location). Our bed and breakfast was closed for the season. Most importantly, we had found a trust worthy caretaker for our flock. We lack any children of our own to which we could assign the task. The changing demographics of our shrinking and aging farm community meant that there are fewer local people of any age upon whom we can rely. Those with any suitable farm experience are too deeply involved in their own operation to care for our flock. Thus, it is no easy task for us to find a caretaker, especially when the commitment was for an entire month.

It was difficult to believe but over the last fourteen years the two of us had never left the farm together for more than a day trip. It is a nice feeling that we had been that satisfied with the life change that brought us to farming. Nonetheless, it was good to get away, especially for so long. With what had begun to seem like endless planning over the previous eight months, the two of us left. On a cold, dreary 5th of November in Green Bay, Wisconsin, we boarded the first of a seemingly long line of aircraft, all but one of which seemed to want to spend excessive amounts of unscheduled time parked on runways somewhere in the U.S. After more than 24 hours of start/stop travel, we arrived exhausted and excited in Auckland, on the dawn of a beautiful, late spring, New Zealand morning.

We had planned our journey so that we could get a brief tour by train, south across the North Island; travel by inter-island ferry to the South Island; proceeding again by train on south to Christchurch. Our Christchurch arrival allowed us three days to explore the city and to attend one of the larger agricultural shows in New Zealand, the Canterbury Agricultural and Pastoral Show. We would then attend the World Congress on Coloured Sheep for a bit over 5 ½ days. Once the Congress was completed, we then hoped to

146

travel through the South Island on our own, making the trip into a true "busman's holiday" by spending much of our non-travel time staying at bed and breakfast style accommodations that just happened to also be sheep farms or at least somehow related to sheep and wool.

I want to be able to share some of our sheep and farm experiences. I suspect that I cannot compress the trip into one chapter; so I will return with further details in the future. I will also refrain from recounting most of the non-farm adventures. (You will need to go elsewhere if you wish to learn of our joys of sitting for hours on a rainy beach watching Fiordland Crested Penguins march up and down the shoreline. The same applies to accounts of silently cruising up the various arms of one of the largest fiords in New Zealand, or discovering that at least one Kea Parrot had attempted to dismantle our rental car.)

Travels to Christchurch

Our initial introduction to sheep in New Zealand was slow in developing. Our early morning trip from the airport into the heart of Auckland did not reveal any sheep, just a few scattering of beef cows in an open area, which otherwise appeared to be in danger of being gobbled up by suburban sprawl. But not to worry, we knew there had to be sheep here since we had already heard them at the airport. There, the passenger service vehicles used to transport people with mobility problems around the airport utilized a unique warning devise. Rather than the constantly "beeping" or "buzzing" heard in other airports, these vehicles emitted "baas". I am sure that I have never before been forced to move out of the way by the mere "baa" of a sheep. But then, this experience was to be a month of firsts.

It was not until our train reached the central section of the North Island that we began to see sheep in large numbers. As expected, they were all white. Nevertheless, it was a delightful sight seeing hundreds of ewes and new born lambs scattered across huge, green pastures that stretched off toward distant volcano cones, partially obscured by clouds. The numbers of sheep became truly impressive the next day after we had crossed onto the South Island and began moving south. Once we passed Kiakoura and entered the northern edge of the Canterbury Plain, the numbers of sheep (and cattle) became staggering. The green rolling fields, which gradually flatten as we moved the 150 kilometers south to Christchurch, seemed to be an endless section of sheep pasture. For the two of us, this area was especially exciting, as it is the area of New Zealand in which most of the larger Corriedale flocks are now found. As the train rumbled closer to Christchurch, we were even treated to the sight of an occasional individual colored sheep and occasionally a small colored flock. The frustration of train travel was that you could not just "pull over" for a closer look.

It was already clear that agriculture was a dominant force in the New Zealand economy. It should be pointed out that our snap shot of New Zealand was in a late spring, which had been (by their standards) unusually cool and excessively wet. Everything, everywhere was green and growing. It was, nevertheless, relatively easy to see where the areas of greater or lesser rainfall would normally occur. Many of our hosts assured us that it was not always this uniformly green.

The A&P Show

Befitting a nation in which agriculture plays a major economic and social role, each area of the country, at some time of

148

the year, has a local A&P show. Each show is the effort of the local Agricultural and Pastoral Association. The shows are, to a degree, a cross between the state and county fairs found in rural areas of the U.S. In a similar fashion, they come in various sizes and with varying emphasis, depending upon the agricultural strengths of a given area. These shows are run under the overall umbrella of the Royal Agricultural Society of New Zealand, a group responsible for the general rules and regulations. In addition to the local A&P shows, there is, each year, a national Royal Show, which is held at one of five different locations. For our trip, we were fortunate that the organizers of the World Congress on Coloured Sheep had purposely scheduled the Congress to dovetail with the Canterbury A&P Show (locally known just as "The Show") held for the three days preceding the Congress. We planned our trip so that we could experience as much of "The Show" as possible.

The Canterbury A&P Show is one of the largest in New Zealand. It attracted over 5000 livestock entries and over 400 on-site trade exhibitors. Located on the outskirts of Christchurch (the largest city on the South Island), it catered to 100,000 people in its three days. We spent the majority of our time in the sheep barns, observing judging and competitions, and just walking amongst the various breed groups, getting a close up look at the entrants and occasionally being able to talk with their owners. Sheep were the largest numbers at "The Show", far outnumbering large groups of beef and dairy cattle, and a smaller number of other species. (Included in the smaller numbers were angora goats and alpaca.) Wednesday, the first day, was devoted to the sheep judging for all breeds. Thursday had additional judging for special awards in some breeds, followed by interbreed judging. Friday was the day of largest public attendance and was, for the sheep, generally a day of intense public scrutiny but little competition. It was also interesting

149

to note that all sheep had to be on site before 8:00AM Wednesday and were not to be removed until after 6:00PM on Friday, which is to say that they were in public view for the entire show.

Nearly each breed group had their own exhibit and judging area. At any given time on Wednesday, sheep breed judging was going on simultaneously in eleven different areas. In all, we counted 18 different breeds being judged. Unfortunately, the concurrent judging meant that we could not watch more than one judging going on at a time. As a result, we chose to settle down on the bleachers next to the Black and Coloured Sheep ring and thus got to see nearly all of their judging Wednesday morning and afternoon.

We do not show our sheep nor can we attend any shows at home, so we have virtually no chance to observe competitions in the United States. As a result, I am not much of an authority on sheep judging and competitions. I am sure I may have missed a few points on which many readers might be better qualified to comment. Nevertheless, I will try to describe the colored sheep judging as best I can. For competition purposes, the sheep were divided into two categories: fine wool breed and strong wool breed. There were no separate competitions for individual breeds. Each group was then judged for rams and for ewes. Within each sex, there were different categories. For the rams the categories were: 1) Ram over 18 months, wooly, 2) Ram over 18 months, shorn, and 3) Ram under 18 months, shorn. For ewes, the categories were: 1) Ewe over 18 months and her suckling lamb(s), wooly, 2) Ewe over 18 months and her suckling lamb(s), shorn, 3) Ewe under 18 months, wooly, and 4) Ewe under 18 months, shorn. The "wooly" sheep should not have been shorn earlier than 16 months prior to the show. "Wooly" rams should not have been shorn later than 8 months prior to the competition and "wooly" ewes not later than 10 months prior to the show. I.e. the "wooly rams" were required to have between 8 and 16 months worth of wool and the "wooly ewes" from 10 to 16 months of wool. From each of these groups were selected a Champion and Reserve Ram and a Champion and Reserve Ewe. Hence, there was a Champion and Reserve Champion for Rams and for Ewes in the fine wool breed and comparable Champion and Reserve Champion for the strong wool breed. From this group was selected a Supreme Coloured Ram and a Supreme Coloured Ewe. Then, between these two, was chosen the Supreme Coloured Sheep.

The judging itself was seemingly an informal, laid back affair. All entrants and judges wore white over-jackets in the ring which lent a professional air to the proceedings. In the case of the colored sheep judging, the white jackets also served to highlight the sheep's colors. It was readily apparent that all the entrants knew each other, at least casually. There was not a sense of intense competition, but rather it seemed as though a good time was being had by all, entrants and judges. Nor was it a competition in which the shepherd "showed" their sheep, i.e. they were not constantly holding and positioning the sheep as one would see in a North American show. Rather, the sheep were loosely restrained while the judge examined them. At one point, the entrants were asked to release the sheep so that the judge could see them as they freely walked and stood independently. The "ewe with her suckling lambs" classes were also shown in exactly that fashion, i.e. their lambs accompanied them into the ring to wander around freely while the dam was slightly restrained. Judging in this case included the lamb(s). As one of the judges pointed out, the lamb(s) had to be "better" than the ewe for the ewe to score well. The judge was looking for a ewe able to improve on her family line rather than stagnate. Many of the ewes in these classes had twins and there was one set of triplets. As might be guessed, it was a noisy, active class.

We were told that judging in the ring environment was relatively uncommon in New Zealand. Apparently in most shows, the sheep in the breed classes are judged in their own pens and then released into the race in front of the pen for the judge to view them walk. In this situation, the judge is helped by two stewards who, in addition to opening and closing the pens, may hold the sheep and then walk them in the race for the judge. In this situation, it is only when the judging reaches the inter-breed classes that the competition moves into a ring and the competitors hold their sheep. In either situation, the judging is obviously much more of the sheep, rather than of the sheep as it is presented by its shepherd.

At the Canterbury Show, when each coloured class was completed, the judge spent time describing in detail, to everyone present, the basis for her decisions, not only for the top ranked sheep, but working all the way down the class. It was a thorough and thoughtful presentation for each group. On the final day of the show, the delegates to the World Congress were invited to a repeat analysis by the judges of the winning animals, with each of the winners being brought into the ring by their owners.

It is also my understanding that the show animals are not to be housed or covered (jacketed), except for when they are in transit to the show and/or for the 48 hours prior to judging. (In our month in New Zealand, we did not see or hear of a single "jacketed" sheep.) Obviously, with the fleece length requirements, especially for the "wooly" categories, one's show preparation may begin and continue through many months prior to the show, in order not to have a damaged or poorly presented fleece, since show time jacketing and trimming do not enter the equation.

We found it interesting that there was seemingly very little emphasis in the colored sheep show on the specific breed of sheep being exhibited. To my knowledge, there are no specific breed associations in New Zealand that recognize colored sheep within the breed. Colored registration is under the umbrella of the Black and Coloured Sheep Breeders Association of New Zealand (BCSBANZ). That registration process, however, seems to be extensive and thorough. Quite obviously, this situation is not unique, since in the U.S. there are few colored breed registries and specific breed registration by an umbrella colored sheep organization is mostly non-existent. Looking at the BCSBANZ flock book, I also got the impression that breed presentation, in and of itself, is not necessarily a major priority for many breeders in New Zealand, but rather what seems most important is the production of a given "quality" and/or "type" of wool, especially with the handcraft market as an emphasis. Often one sees a flock described as being based upon a specific breed, with the addition of a different breed ram to somehow alter the fleeces in a specific direction (e.g. to add softness or color patterns). Thus, one will see a description of a flock as being, "a Romney type" group.

When the Black and Coloured sheep judging was completed, we had an opportunity to move on to the Corriedale judging (fortuitously, just opposite the colored sheep in the show barn). The Corriedale class was one of the largest (if not the largest) at the show. It was our understanding that this was traditionally one of the largest and most important Corriedale gatherings of any show in New Zealand. The breed was originally developed in the Canterbury region and is ideally suited to its geography and climate. The vast majority of the New Zealand Corriedales still come from the greater Canterbury region. We had little chance to see much of their judging as it was nearly over when we arrived. We, therefore, spent much of our time walking up and down the races, looking at the rams and ewes in their pens, and getting to talk with several of the exhibitors. Some of our observations follow here.

We saw Corriedale rams and ewes with a tremendous amount of wool, both in terms of staple length, density, and overall covering/yield. Their wool tended to have a very pronounced crimp from our perspective. We felt that most of the fleeces we saw were perhaps not as fine as our personal flock's or their Australian counterparts. The rams were usually 30+ microns and the ewes 28 to 29 microns. There was definitely an emphasis on consistent, overall wool quality and yield, with little difference observed in wool from top to bottom and end to end. The wool toward their bellies appeared to be nearly the same length and quality as much higher on their frames. (We saw an even greater emphasis of this in Merino rams we saw being judged. The judge had all the rams tipped up so that he could examine their belly wool, stating that good quality bellies served to enhance the overall package of wool that the sheep presented.) The overall size of rams and ewes was perhaps smaller than the "modern" U.S. Corriedale. There was seemingly little concern for breeding a tall, long legged breed as there currently seems to be here.

The Corriedale judging was broken into a large number of categories, both as individuals and as groups. We regretted not being able to observe this judging. An especially interesting judging group (which appeared to be unique to the Corriedale competition) involved wool judging before and after shearing, with consideration given to judging the sheep that provided the fleece. As part of this competition the shorn fleece underwent a micron analysis. When completed, the fleeces were displayed above the pen of the sheep that sourced the wool. The display was accompanied by a summary of each micron-analysis. It would have been interesting to have seen at least a portion of this contest as it was in progress.

We left the A&P Show on its final day, overall very happy and impressed with the show. We did not get to witness any of the wool competition, as judging was already completed before the show opened. We saw some beautiful fleeces (colored and white) both on and off the sheep. The number of moorit sheep and fleeces in the competition (in both the fine and strong categories) was impressive. There were a couple of Corriedales (both colored and white) that we would have been thrilled to take home with us. But somehow we felt that three weeks in a small car would have been a bit hard on both us and the sheep. The prospect of trying to explain our acquisitions to customs, immigration and security in the U.S. was at bit too daunting! But, we did learn a good bit more about the export/import of semen for future artificial insemination in our flock. The export/import procedures are a process that is more complicated than we imagined it to be, but we at least have some leads to pursue.

We managed to get a short peek at the shearing and wool handling competitions. Herding dog trials got left out due to lack of time, much to our regret. We did have some interesting discussion with some of the trade exhibitors, especially one selling some of the slickest looking sheep handling/sorting systems imaginable. It also was a bit too much to fit in a suit case, but true to the salesman he was, the farmer who made the system assured us that it would be "no problem to ship the unit to us in the U.S." Lastly, we got through the three days of the show without talking to a single unpleasant, unfriendly person. Every exhibitor was more than happy to talk with us, even as they were preparing for their next competition. Even conversing with other spectators proved rewarding. As with the rest of the trip, everyone we met was extremely open, warm and friendly.

The Congress

After having immersed ourselves in sheep and wool for three straight days at the A&P Show, moving onto the World Congress on Coloured Sheep was an exercise of just turning up intensities a few more notches. I will not attempt to describe the proceedings in detail as such an attempt by me will do them injustice. It was a very professionally organized conference, especially when one considers the volunteer nature of all the people responsible. The written summary of all the presentations includes additional articles and a large selection of color photos. It is an especially impressive feat. Special praise should be given Roger Lundie and Elspeth Wilkinson for their masterful job of editing the book, The World of Coloured Sheep[1], which is available in the U.S. through Black Sheep Newsletter[2]. It does a much better job than I can ever hope to in detailing each of the presentations. I highly recommend reading it.

The Congress was held in Christchurch. We arrived just in time to enjoy the late spring flowers blooming in parks and private gardens everywhere. Especially impressive were the hundreds of rhododendrons and azaleas in full bloom. Christchurch is an attractive city if one must be in an urban environment. The Congress was attended by about 250 individuals. The make up of the group was, as to be expected, mainly from New Zealand and Australia, with additional participants from Japan, Brazil, South Africa, Sweden, England, Wales, Scotland, Canada and the United States. The Congress afforded the two of us the opportunity to

---

[1] Lundie, Roger S. & Wilkinson, Elspeth J., (Ed) (2004) The World of Coloured Sheep, The Black and Coloured Sheep Breeders Association of New Zealand
[2] Black Sheep Newsletter, Scappoose, OR

156

renew a few old friendships, make new acquaintances, and in some cases put a face and personality to names we have often seen but never met.

Congress sessions were especially intense. It was at times difficult to digest the shear amount of material that was presented in such a short space of time. It was good planning to have many of our afternoons broken up by outdoor presentations involving live sheep and lambs, especially as they related to colored genetics.

As is the case in many such conferences, some of the most lasting memories and lessons are the result of the personal acquaintances one makes. On one afternoon, while Gretchen immersed herself in sessions related to woolcrafts, I was spirited off by some of my fellow Kiwi shepherds to see a couple of sheep farms. The first operation was part of Helen Heddell's "Hiltonblacks" farm on the Canterbury Plains, just northwest of Christchurch in Swannanoa. Helen raises colored Merinos and New Zealand Halfbreds, in addition to a large flock of white sheep and beef cattle. Helen brought some of her colored sheep to the Canterbury A&P Show and had won the Supreme Black and Coloured Ram (a truly impressive Merino ram) and Supreme Black and Coloured Ewe award (a lovely young Merino ewe). The ewe was chosen Supreme Black and Coloured Sheep. It was good to be able to see the sheep, with whom we had become familiar at the A&P Show, out in their own farm environment. The farm was characterized by large flat, irrigated pastures, each of which had extensive planted wind breaks, very characteristic of this area.

After "Hiltonblacks", our little group headed north about 85 kilometers, into the heart of Corriedale territory, past Waikari, to Bev Forrester's farm "Black Hills". Bev has a colored flock of Merino, Corriedale, New Zealand Halfbred and Romney, in addition to a large white Corriedale flock and beef cattle. This far

157

north, the land has started to rise dramatically from the Canterbury plains, becoming increasingly mountainous. Much of the land is in pasture on relatively steep terrain. The majority of the 2250 acres has been cleared or is being cleared of the brush and scrub. The sheep have the run of extremely large pastures. In addition to the agricultural enterprise, Bev has restored the original pioneer cottage and stable on the farm. Besides offering historical tours of the property, she runs a fibre gallery in the stable, selling handknit and handwoven products made from the colored wool she produces. Here again we got to see some of the show sheep back on pasture near the stable. This group included a fine colored ewe with triplets who by the time the show had ended looked very much depleted. It was amazing to see how well she had rebounded in just a few days back at home on pasture. Bev also took us on a tour of the larger hill pastures to view the white Corriedales and finally to see the grand view from the highest elevation on the property (from which one could see all the way to the Pacific Ocean to the east, and north and west into the heart of the mountain ranges which, at the time, were getting a fresh layer of snow). It was an afternoon not on the official Congress schedule, but it was worth each minute to see, as one of the shepherds remarked, the "real New Zealand".

*****

With the Congress over we are now left with a tremendous amount of information and ideas. I need to wait a bit to let much of the information from the Congress ferment. What remains are lots of pleasant memories and new friends. We would like to add our deep, heart felt thanks and appreciation to all the members of the Black and Coloured Sheep Breeders Association of New Zealand

158

for putting together such a wonderful Congress, and especially for welcoming us with such warm and friendly hearts.

I will return with an account of our post-Congress travels to visit sheep farms across the South Island of New Zealand.

Adventures in New Zealand

In November of 2004, we traveled from Wisconsin to New Zealand to attend the Sixth World Congress on Coloured Sheep. When last I wrote, I recounted some of our experiences related to the Congress. This time around I would like to describe some of the sheep related adventures we had once the Congress was completed. Gretchen and I had been lucky to arrange for a month's worth of "farm sitting" for the trip. The extra time allowed us to travel around and through a fairly large portion of the South Island of New Zealand. It was our hope to combine our interest in sheep with a fairly leisurely getaway from our own flock for the first time in close to15 years.

In New Zealand, one is able to find a goodly number of farms that offer some sort of "bed and breakfast" type accommodations. We specifically sought out such locations, since it mirrored our own experience of both raising sheep and hosting a bed and breakfast. Of the nine different places in which we stayed over a nearly two week period, five were either operating farms which included sheep or were in some fashion related to the New Zealand sheep and wool industry, past or present. Each of these accommodations were, in one way or another, connected to what would be considered the mainstream sheep industry of New Zealand -- white sheep and wool. The only reason that the other four stays were not sheep related was that they were in locations in which either no farm stays could be found or were in locations which lacked sheep. Unfortunately, from our point of view, none of the farm stays included sheep purposely kept for their color. On much of the South Island, we were in awe of the large numbers of sheep being raised in a diversity of habitats and environments.

We were in New Zealand in their mid to late spring, which meant that we had one of the greenest overall images one can get of the South Island. Strange weather patterns, however, seemed to have a way of following us from our Wisconsin home. Their spring was later, wetter and colder than normal. Snow seemed to fall nearly every day in the higher elevations of the Southern Alps, and on two occasions managed to move down to the lower elevations over night. It was a frustrating time for the farmers we met. Their pastures were green and ready to explode, but the lack of warmth and sunshine was severely limiting the growth of the forage, at a time when lactating ewes and growing lambs needed it most. Nearly all of the New Zealand livestock industry (sheep, dairy, beef and deer) is pasture based, so the delay or lack of a spring flush of forage growth was trying for many farmers. Only when there was too much forage to keep up with grazing would there be any mechanical harvesting. When the forage is cut, much of it ends up as baleage, rather than the drier cured bales to which we are more accustomed. Many of the farms rely on custom harvesters, rather than making their own hay. The late spring also meant that the custom harvest crews were on hold. We could only hope for everyone concerned that the sun would come out and the temperature rise a bit.

Paua Bay and the Banks Peninsula

For our travels we rented a car. Generally, we faired well with our first experience of driving on the left hand side of the road. However, with the frequent rain storms, I often had need of the windshield wipers. The controls for the wipers were where I was used to having the turn signals. We had frequent fits of frustration and laughter as I tried to signal our turns with the wipers or fight off the rain with the directional signals. New Zealand survived our assault on their roads. For the most part, we aimed for the less traveled roads, which often were gravel, one lane tracks. Our major traffic problems were generally related to meeting an on-coming stock truck on these back roads and then having to seek a suitable space to squeeze past each other.

Our first destination, after leaving the World Congress in Christchurch, was the Banks Peninsula. An easy day's drive from Christchurch, the peninsula is formed by two extinct volcano cones which jump out of the Pacific Ocean at the edge of the otherwise very flat Canterbury Plain. The peninsula lacks much in the way of level ground, but rather is mostly steep slopes, descending into ravines, which in turn rapidly drop toward the ocean. At one time most of the peninsula was forested, but European settlement resulted in extensive timbering. Much of the land is now either in scrub or is extensive pasture land for sheep, cattle and/or deer.

We spent two days at Sue and Murray Johns' Paua Bay Farm. It is located on the extreme eastern edge of the peninsula, tucked between the upper extremes of two volcanic ridges which dramatically drop to the Pacific Ocean. From the farm house, which is surrounded by a lovely garden, one can look across the valley to sheep pastures on the opposite ridge, which extends for a couple of miles down the step slope to the beach. They farm a couple of thousand acres in two separate properties, raising some cattle and deer, but primarily sheep. Their flock is about 2000 crossbred white ewes. The foundation of their current flock is Perendale (a dual-purpose breed, developed in New Zealand through the interbreeding of Romney and Cheviot sheep). Their fleeces are bulky and of low luster, with an open staple. The breed is popular in hill farming as the sheep do well on poor pasture. Of late, they have been crossing their ewes with East Friesen (to increase size and milk production, Finn (to increase prolificy) and Texel (to improve the meat carcass). Theirs was the first of many farms where we saw this extensive amount of cross breeding (not always the same combination and mix of breeds). In each case, their primary goals were to improve the number of the lambs they produced, maximize their rate of growth and improve their slaughter qualities. As elsewhere, the general wool market in New Zealand is struggling, while the slaughter market is increasingly the backbone of most sheep farmer's income. It was interesting to see the fleeces from these composites, both shorn and on the ewes. It was easy to see the influences of the diverse breeds and equally easy to see that consistency in fleece type and quality was not a paramount concern.

On the second day of our stay, Gretchen and I hiked the track down through the pastures to the beach. The pastures are often too steep for any type of mechanical intervention; the sheep and their rotation through the large paddocks are the major source of pasture maintenance. The pastures were a mix of cocksfoot (orchard grass to us North Americans), rye grass and white clover. Thistles can be a problem (here and elsewhere). In this situation, their only control seems to be spot spraying.

Our hike was a delight. It proved to be one of our few warm, sunny days. We "shared" the grassy slopes with the skitterish ewes and their lambs -- an upper paddock was devoted to older ewes and their lambs, while the lowest paddock, which stretched to the beach, was occupied by hogget ewes and their lambs. The rocky beach was also occupied by a few fur seals and was littered with paua (abalone) shells. On the way back up, we stopped for our lunch on a grassy ridge, overlooking "our private" beach, enjoying local cheese, bread and wine, along with a magnificent view. Our lunch time entertainment was watching the neighbor farmer mustering sheep off the face of what seemed an almost vertical ridge opposite us. The priceless value of a good Border Collie was never more evident. It would otherwise have been impossible to gather the ewes and lambs scattered on this terrain.

Like most of the sheep operations we saw, the physical plant of Paua Bay Farm was minimal, primarily consisting of a shearing shed and small equipment storage area. Of much greater significance is the mile upon mile of fencing that encloses the various pastures everywhere. Fence building in New Zealand is both an art, skill, and often a physical feat. As in many such operations, available labor was at a minimum. There was just Murray, Sue, and their son who had just decided to join in the business. The only major outside help is from the contracted shearing crew. The farm was also a diversified operation. The Johns' have been running their small bed and breakfast for 14 years (a long time for anyone in this type of business!). Interestingly, they are currently experimenting with an additional spin-off from the tourist <u>and</u> agricultural side of sheep farming. Murray is now scheduling shearing demonstrations for which admission is charged

Our memories of the stay at Paua Bay are all very fond. We shared some lovely meals and conversation with Sue and Murray, especially a wonderful lamb dinner on our last night. It was hard to match such an environment. Falling asleep with the quiet, occasional baaing of sheep near the house, mixed with the more distant roar of the Pacific Ocean are memories which will remain with us for a long while.

The Southern Alps and Arthur's Pass

Our next stop was a leisurely day's drive west from the Banks Peninsula. We came down off the steep ridges to the flat, irrigated farm lands of the central Canterbury Plain and then proceeded further north and west, eventually onto the eastern side of the Southern Alps. We stayed at the Wilderness Lodge, adjacent to Arthur's Pass National Park. The lodge is situated on a 7000 acre sheep and cattle station. When the station was sold to its current owners, Gerry McSweeney and Anne Sanders, they chose to create an upscale eco-tourism lodge in the middle of the property. They purposely removed from production some of the more marginal pasture land on the mountain slopes and instead devoted the higher elevations to wilderness restoration/preservation. With a very strong emphasis on environmental restoration and preservation, the staff conducts guided hikes, kayaking and similar activities for their guests. What makes the lodge especially unique is that they also continue to operate the sheep and cattle station, again attempting to employ environmentally friendly grazing and farming practices. As an integral part of the program for their guests, they offer tours of the farm. Gerry is able to educate the guests about the details of raising sheep and cattle in the high country, while usually providing tours of the shearing shed, including shearing demonstrations.

Their flock consists of about 3000 super-fine white Merino ewes. As might be guessed, they are concentrating on the production of a fine wool product and secondarily on lamb for slaughter. The wool shed is immaculate. It is set up with shearing stands for either five machine shearers <u>or</u> five blade shearers. The majority of their super-fine Merinos are blade sheared. Blade shearing is done, in part, to leave a longer wool residue on the breeding ewes in order to afford them more protection from the sub-alpine weather they often encounter. In the shed, they have installed a micron screener (attached to a computer) so that as soon as a fleece is removed from the sheep it can immediately be tested for fineness. In this fashion, each ewe's fleece is measured, weighed and recorded at each shearing. This procedure also permits the fleeces, immediately after shearing, to be sorted and baled into lots based on a very specific micron measurement of the fleece. Their ewes are producing fleeces with a range from 16 to 18 microns. These fleeces are sold to top-end clothing manufacturers in Europe at a price around $12NZ/lb (a price well above the New Zealand fine wool price at the time). Clearly, this emphasis on wool quality and production was different than we were to see elsewhere.

Ironically, the production of expensive, super-fine wool also comes with a major downside. These Merino ewes and rams tend to be small, with a rather meager meat carcass, so they are not able to capitalize on the robust lamb market that New Zealand is currently experiencing. In the past, the Wilderness Lodge farm had sought to improve their market lamb size by breeding their ewes to larger, black-faced, Suffolk-type rams. The resulting lambs of this terminal cross do have an improved size, but the Merino ewes tend to have difficulty with delivery due to the much larger birth size of the lambs. As a result, they are now experimenting with Dohne Merino rams. This breed is a polled South African Merino with much greater size than the super-fine Merino. So far, they have not experienced the degree of delivery problems associated with the black-faced rams. They are also hopeful that the offsprings' fleeces may remain fine enough to command a higher premium than those of their current terminal sired lambs.

166

Wilderness Lodge's entire operation is impressive. They do a wonderful job of positively presenting the sheep and wool industry to the tourist. Included in the Lodge lobby is a spinning wheel and a good sample of rovings from Merinos and other breeds. Normally, Anne is available in the evening for spinning demonstrations. She was absent during the evenings of our stay, but Gretchen received permission to try the wheel and some of the samples one afternoon. Before she knew it, she had become the demonstrator. (By her account the Merino was a lovely fleece and the wheel just so-so. She was, however, very happy to get back to our Corriedale wools and her Lendrum spinning wheel!) Gerry and Anne are to be congratulated for presenting sheep and wool in such a positive, public spotlight. One cannot but feel excited by the sight of their beautiful flock, grazing on broad, flat pastures in the Waimakariri River valley, flanked on three sides by snow capped mountains.

Wanaka

After spending a few very exciting days on the south west coast watching penguins in the rain, we moved on for a single day stay at a lodge specifically built as a bed and breakfast, on a farm just outside of Wanaka. Wanaka is farther south, on the eastern side of the Southern Alps, in an area which, quite obviously, does not receive as much rain as many of the areas we visited. The farm on which the Riverrun Lodge is located is a diverse product operation, raising a large acreage of grain (this year, primarily wheat) and a crossbred flock of about 2500 lambs that they purchase and finish on pasture prior to marketing each year. The sheep were at a remote location at the time, so we saw none of them. During the winter, they take in dry cows to graze their pastures, prior to returning them to their owners for calving. They also maintain a small flock of mixed breed colored sheep, whose sole purpose seems to be maintaining, through grazing, some of the perimeter areas adjacent to their crop lands. We finished our evening in Wanaka with another good meal featuring lamb and locally produced wine. It was a pleasant evening of conversation with our hosts and an Australian couple from Sydney. Nevertheless, it was more of a business type environment, rather than the feel of being with fellow shepherds.

The Southern Coast

In a couple of days, we got back into a full fledged sheep environment on the extreme south coast. We spent a night with Ann and Donald McKenzie at Greenbush, their farm and home outside of Fortrose, east of Invercargill. It is an area of beautiful rolling coastal pasture land. Ann and Donald are now retired from sheep farming, having turned the farm over to one of their sons. Their second son also farms just down the road. Donald still spends much of his time helping with the farm and flock. Ann's and Donald's home is the original farm house of his ancestors going back to 1865. It is a lovely old place that reflects the grandeur of the heyday of New Zealand sheep farming. The home is nestled in a spectacular garden surrounded by large mature trees. One hardly knows that the shearing shed is right next door.

Donald took us on a farm tour that evening. The farm has the advantage of being mechanically accessible. As such, they are able to clip, fertilize and renovate the pastures as they require. The result is truly beautiful rolling pasture, some of which stretch right out to the beach and the Southern Ocean. The McKenzie's farm about 1000 acres in all. The flock, like so many others, is a composite one. In this case it is Romney and Perendale based. They

168

currently are using Texel rams as terminal sires, again with the purpose of producing a growthy, meaty lamb for market. As with so many other places we saw, quality wool production has become secondary, due to market conditions. We got the sense that Donald was hoping (dreaming?) that the market for strong wool might rebound someday. They were managing their breeding program such that they would be able to rapidly re-emphasize the Romney wool component if it again became important. One had the sense that wool was the component that the McKenzies loved most about sheep and that they were hoping for a rebirth of its importance. This love of wool is also reflected in Ann's interest in spinning, knitting and weaving.

On to Corriedale

With thoughts of the past glories of the wool industry still fresh in our minds, it was somehow fitting that we would spend our last night in New Zealand at Tokarahi Homestead. Being Corriedale breeders, we felt that it would be fitting if we could visit the birthplace of "our" breed. We located "Corriedale" on a map, northwest from the east coast city of Oamaru. We found the spot, in a lovely farming area. The development of Corriedales had first begun by Mr. James Little at a sheep station called "Corriedale" in 1868. In our mind, we had pictured a solid stone estate from the era, still standing on the spot. It was not to be. Rather, a stone marker from 1940 and a directional arrow are all that appears to stand there now. Up a gravel track, roughly where the arrow pointed, we heard and eventually located a large flock of sheep. The ewes had obviously just been separated from their lambs. The clamor between the two pastures was intense. No shepherds were in sight, but the sheep appeared to be Corriedale. It was very easy to fantasize that this was a scene from one hundred years previous with some of the early generations of James Little's sheep creations.

Up the road, perhaps 15 kilometers, is the Tokarahi Homestead. It is the building we fantasized for "Corriedale". A beautiful cut stone house, built in the era when Corriedales were created, it was once the home of one of New Zealand's wealthy wool barons. During more recent times, the building and grounds fell into great disrepair. Luckily, it was saved by Mike and Lyn Gray, who spent years lovingly and painstakingly restoring the place to its former glory. While they do not farm and were only able to buy the house and a couple acres on which it stands, they have a couple of pet wethers ("William" and "the Lone Ranger"). All around them is a large, active Corriedale sheep farm.

It was a memorable final night for us. Our hosts epitomized the friendly, open nature of all the New Zealanders we met. We had a wonderful dinner (yes, once again it was lamb!), pleasant conversation and, lastly, a chance to look at a finally clear southern night sky through Mike's telescope. Mike and Lyn are going to have to sell their place for personal reasons. It was an easy extrapolation for us to imagine ourselves, in that location, if only it had also included the farmland. It would also mean having to leave our American Corriedales behind, which is a step we are not prepared to make as yet.

It was good to return home, to be greeted by our own flock and to have them again come to greet us, rather than to turn tail and slip away as did all the large farm flocks we visited in New Zealand. Just as in this county, most sheep people we met in New Zealand cannot seem to fathom the notion of raising a flock that is color based. The notion of such a flock even partially supporting a way of life is alien. For this reason we value the friendships that we have here and that we have made in New Zealand, friendships that are ground on a common understanding of the uniqueness of raising colored sheep. Because of events like the World Congress on Coloured Sheep, we are able to expand that friendship. Thanks to all of the Kiwi's who were so warm in opening their homes and hearts to us!

******

It is now difficult to imagine that it was not that long ago that we were in New Zealand. Shearing is just finished and none to soon. Our lambs started arriving a few days after shearing and, just as predictably, the weather has turned cold with lots of snow. So we are back into the frustratingly joyous period of lambing. The arrival of healthy twin black lambs from a favorite ewe, early on a frosty late February morning makes it all worthwhile (and good to be home)!

Observations from the
End of the Parade

Today, the ewes and their lambs are just over half finished with the season's first grazing of the pasture we have come to call "The Orchard". Even though the now departed cherry trees seem like ancient history to us, their one-time location will always be "The Orchard". It is one of my favorite pastures, for a number of reasons. Its fertility is excellent and the grasses and clovers that grow there always seem to do well. For our farm, it is high, open ground and therefore the breezes tend to keep the mosquitoes and black flies at bay. For that, both sheep and shepherds are always grateful! The overall layout of this pasture has made it one of the easiest of our pastures in which to make the daily temporary electric fencing move. It is also closer to the house than most of our pastures, which means that the flock is always within easy view. From a security standpoint, the close proximity is a blessing. Best of all, it lets us enjoy the flock, close up, throughout the day.

Getting the flock out to "The Orchard" each morning is, however, one of the longer walks that the sheep must make from the barn. The walk from the barn, down the lane along the old stone fence, through the first pasture, up the rise to the high ground and then back to the far end where they are now grazing is perhaps a quarter of a mile. Even though the adults now know the pasture well, it sometime seems as if they are not sure it will be there each morning, until they get to the top of the rise and can see it. Once the destination is in view, the rather halting parade picks up ample speed. By the time I catch up, most of the flock is eagerly grazing and the rest is looking for "the perfect spot" to start their breakfast. We even bring the swallows with us from the barn. The flock's arrival in the new pasture kicks up the bugs that provide the swallows with their breakfast. Someday, I would like to get there ahead of the flock just to watch from the front as the whole procession of sheep and birds arrive. Unfortunately, I am fated to bring up the rear, closing gates and making sure that everyone arrives alright. In addition, I walk a lot slower than any healthy sheep.

So far this year, it has been more of a struggle to get an individual ewe to take charge of the procession to and from the barn, no matter which pasture is our destination. It has been a few years now since we last had a good, consistent "leader". In the good years, a special ewe is almost always first out the barn door and at the head of the parade to pasture, setting a good but steady pace. The need for a leader may even have a bearing on our culling selection this year. One of the few ewes to exhibit any willingness to assume a leadership role is Marvelous. She is a seemingly unlikely candidate. A four year old ewe, she never exhibits signs of dominance over any of the flock. She is friendly, a good mother and comes from a family line noted for being quiet and unassuming. Unfortunately, she may have suffered a bit of loss in milk production with a case of mastitis this spring. We were able to catch it early and we have (hopefully) minimized the damage. Nevertheless her milk production did drop, based upon the performance of her twin lambs. With that type of difficulty, she managed to get on our list of potential adult ewe culls. As we approach weaning time for the lambs, in preparation for selling most of them, we also start making the personally difficult decisions as to which adults may need to also go. In Marvelous' case, she may just have assured herself a place in the flock for a while longer, even if we do not breed her again, at least if she is willing to continue to lead the parade each day.

We do tend to keep a few of our ewes who, for whatever reason, we no longer breed. The criteria for staying are relatively simple and not based on too many business considerations. The ewe must still be able to walk comfortably to and from the pastures everyday. She must still produce a good to excellent fleece. (She would not have been originally retained for breeding without a very good fleece.) She must have a good, friendly personality. We can forgive the standoffish or flighty ewe if she is consistently producing good wool and nice, robust lambs. But if she is removed from lamb production, "spooky" will not be sticking around unless she becomes a bit friendly.

Interestingly, the daily parade to and from the barn is often one of the best places for analyzing the performance of ewes and lambs. Since I am nearly always bringing up the rear, I become familiar with the individuals who daily tend to be at the back of the flock. Just as there are leaders, there are "dignified ladies" who would just as soon avoid the rush. The presence of a different ewe near the end of the line is almost a guarantee that something is wrong with her, either due to illness or injury. In some situations a different behavior, for one of the "old ladies" who is usually at the end, can also be telling. Two of our oldest retired ewes, who have helped me bring up the rear for a number of years, are noticeably slowing down this spring. Both Enchilada and her oldest daughter, Finale', have been slowed by arthritis. This spring the arthritis has gotten worse for both. It also means, in all likelihood, that they are not standing as much and therefore grazing less during the day. If this proves to be true, they are going to slowly loose condition at a time when they should be filling out on lush pasture. In a similar vein, Iona and one of her daughters, Kona, are showing up regularly at the end of the line and also moving slower this spring. They have a family history of mobility problems related to their hips. They have outlived the rest of the family without showing the standard symptoms...at least until this spring. I know that when I can keep up with them on our walks to and from pasture that they have slowed down significantly.

There are others in the "retired" group that blend into the rest of the flock. It takes a good eye to spot them (a bit over-conditioned perhaps, heavier fleece, no active udder, etc.). They may stay that way for a long while. I am always glad when the back end of the parade is just as tightly packed as the front. It means that I am much farther behind, but it also means that everyone is outwardly healthy and strong.

I have managed to record my thoughts this afternoon under good conditions. The flock is in "The Orchard", just beyond where I sit to write. It takes twice as long to write, because I find that I am frequently stopping just to watch and enjoy the flock. They seem contented with grazing. It is cool and overcast (good "sheep weather"). Tomorrow, the sun should be out and I will, hopefully, be on the tractor starting to cut hay. At least, for this afternoon, I have managed to spend more time with the flock.

Long Term Memory

The current project on my loom is a wool blanket, woven with wool from our flock. Winter is my time for serious weaving. The sheep are off the pastures, as there is nothing to graze. Snow is blowing and drifting past the window and it is cold (colder than normal for December). It is a fine time to be inside and enjoying some of the fruits of the flock's fiber output. It is strange that under these conditions that I can be distracted, but once I am into the rhythm of the treadling patterns on the loom I suppose that my mind will, on occasion, wander off somewhere, leaving the rest of me behind. Perhaps a memory gets in the way from time to time. How else can I explain the mistake in the weaving that I discovered a while ago? Mind and body now re-united, I have just about finished "unweaving" a foot or so of the blanket, back to the mistake. As I removed length after length of weft yarn, I was struck by the memory possessed by the two ply yarn with which I had been weaving. After a day or so in the partially woven blanket, the yarn reflects a strong memory of the pattern it once helped to create. Even after the yarn had been removed, it was easy to see where a section was woven in plain weave and then switched to a section of waffle weave, the two patterns alternating in stripes up the warp. The once plain weave yarn is tightly crinkled from over/under repetition, while the waffle weave yarn's impression is much less defined and dramatic. Only when I eventually rewove the area, with this same yarn, did these memories disappear, presumably replaced by new impressions.

The yarn's rather temporary memory started me to think about related, more permanent memories associated with the sheep which had produced the wool for the yarn. I am often struck by how long a sheep can remember places, objects, people or events. We often joke about the brain damage the rams inflict upon each other with their occasional bouts of serious head butting. Repeated head trauma does not seem to totally diminish their memories. This phenomenon was apparent this fall, as we readied the rams for the breeding season. A day or so before Gretchen and I put each ram with a group of ewes, we put the marking harnesses on the rams. It gives the harness a chance to settle into the ram's jacket and fleece, while allowing the ram to become accustomed to its feel before the chaos and great excitement of the ram's first day with "the girls". Putting a harness on a ram lamb is, for us, a major production. It obviously feels strange to the young lad, both going on and once in place. Luckily our ram lambs are only, at best, two thirds grown, so restraining each one is not quite as difficult as a full grown Corriedale ram would be. This year's two ram lambs, Stud Muffin and Sam the Ram, were no exception. They refused to hold still, fighting the process the entire time that the harnesses were being fitted. Once a harness was in place, it took at least an hour for each ram lamb to settle into the sensation of having the harness wrapped around them.

When we were able to turn our attention to the adult rams, everything changed. All of the adult rams were quietly and calmly serene. If they were as calm for the rest of the year's chores, life would be very boring (and a lot easier)! They obviously remember the significance of the harness and the memories of good things to come. It takes us much less time to get harnesses on six adult rams than it took for the two ram lambs. The only slight exception was Rhett, who was a little wiggly. Even that behavior makes perfectly good sense as he was a lamb a year ago and has fewer memories to fall back upon. For the other five adults, there was no dancing around. Once outfitted into their harnesses, they were relaxed, at ease and ready to go. Of our adult rams, we had decided to give Mercury and Ironsides the year off. If one can read disappointment on a crinkled up face of a ram, it was written all over theirs when it became obvious that they were not to be fitted for harnesses. Their memories were as strong as their fall odor and their disappointment.

Omar, in harness, strutting his stuff

Later this fall, the ewes gave us another lesson in long term memory. During breeding, all of the adult ewes that we decide to try to get bred spend a little over two heat cycles, i.e. a little over five weeks, in small groups, each with their own "private" ram, each group on a different pasture. It is the one time of the year in which they cannot and do not come back to the barn at night. In a sense, they are away from home for over a month. When it is time to bring them back home and to put the rams back into their bachelor quarters, it is a hectic time for us, mixed with a lot of relief when we know that we have managed to again avoid any predator damage. For the ewes it is a social time, when they get reacquainted with each other and often reunite family groups.

In the barn, we have one automatically filled waterer that is capable of taking care of at least as many as 130 ewes and their lambs. The same unit has seen lots of heavy use for almost 15 years. We have been extremely pleased with its performance as it has done yeoman service. This year it finally was starting to show its age. While the ewes were on pasture fulltime for breeding, we took advantage of the waterer's down time to replace it with a new one. Outwardly the new unit looks virtually the same as the old one, except that it is shiny rather than dull and tarnished. Mechanically it is slightly different, making a new, yet quiet noise while filling. One would have thought that the new waterer also had flashing strobe lights attached to it on the day we brought the ewes back to the barn. The ewes remembered the old waterer and this one was not it! As a general rule, the older the ewe, the longer it took for her to get up enough courage to approach the new waterer, no matter how thirsty she might have been and no matter how many others were already drinking from it. The ewe lambs were the first to use it and to not be spooked by it. After a few days, it had become comical to watch the dignified older ewes screwing up enough courage to approach the waterer ... but only if it was not in the process of refilling! It was amazing how a seemingly minor alteration in the ewes' daily environment could appear to be so obvious and so ingrained in their memories.

By this time I am sure that the flock was wondering "What else can they try to do to us?" They at least had some warning about "The Chickens" and the lambs already knew about "The Cat". This summer we had begun our first adventures with a flock of laying hens. We started a group of Barred Rock hens from day old chicks. By mid-summer they were outside in their movable hen house and pen, learning to graze. The ultimate goal may be to have the chickens following the sheep from one pasture to the next, in the hope that the chickens will help reduce the potential parasite load for the sheep. This year, however, the chickens were too young, too small and too few to keep up with the daily movement of the sheep. Instead, they grazed (and fertilized) the lawn near the house and barn. They did on occasion come close to crossing paths with the sheep. If the sheep were in a pasture next to the house, they could be seen watching the chickens through the fence with great curiosity. The chickens provided them with great amusement on these occasions, but always from a distance. (I have yet to figure out the chickens' reaction. They are weird creatures compared to the sheep, but we have enjoyed them!)

Once breeding was over, the chickens proved to be a much more serious concern for the sheep. For the winter, we had decided to move the hen house into the newer wing of the barn. There they were to share winter quarters with the rams. As we brought each breeding group back from their pastures into the barn, we would do a number of chores with them: changing jackets, trimming hooves, worming and, of course, separating the ewes from their ram. Once the ram had his harness removed, it was time to move back into the familiar confines of the "new barn", just like they have done for a number of years. It was not, however, business as usual as far as their memories told them. There were these strange, cackling, feathered creatures, right next to their pen. It was not necessary to ask which was more frightening: a pen of eight hens or the prospect of head butting seven other grown rams to re-establish an order of dominance. For the better part of a day, the chickens were the main interest. Without fail, the rams chose to avoid the chickens as much as possible and forgot about combat. The ram's memory of the barn did not include chickens. It has taken a while but "the boys" are now content with their new neighbors.

Thankfully, the ewes did not have to contend with the new waterer and the chickens in their portion of the barn. However, upon their returning to the barn after breeding, the entire adult flock needed to adjust to one more change. It seems that just about every fall someone dumps one or more cats on our farm. We live in an area populated with more than its share of seasonal homes. Early in the year, at least a few of these seasonal residents bring a cat with them to clean the mice out of their summer homes. Every fall, when these same neighbors begin to close up their houses and head south for the winter, the most "convenient" solution for the cat is to leave it at a farm (unannounced). I suppose that their logic is that a farm should be a good home for their temporary mouse trap. So every year we must put up with a terribly nervous, high strung, uninvited cat lurking around the barn. There are usually lots of mice to catch and two humans to avoid at all costs. When we accidentally corner the poor creature, we have to deal with a virtual explosion as the cat rushes into hiding. Unfortunately, the same behavior often occurs when the cat and the sheep confront each other unexpectedly. In this situation, the result is two explosions: one small and feline, the other massive and ovine. It becomes an unpleasant environment for sheep, shepherds and cat. Either the cat suddenly disappears on its own, usually within a month or so, or we eventually feel the need to trap the cat and find it a better home. In addition to the heighten anxiety level the cat brings, we are also concerned about the possible bio-hazard. A cat with unknown background is an easy vector for toxoplasmosis, a disease which can cause abortions in pregnant sheep (and humans). It must be noted that the control of the mouse population is nice. Yet despite a reduction in the mouse population, we do not feel comfortable with the anxiety level that our sheep begin to show. It has come to the point that the sheep almost seem to expect (and dread) the presence of a spooky cat in the barn at the end of breeding season.

This fall began exactly like most others in terms of the appearance of another cat. It showed up about the time we put the ewes in with the rams and it decided to hang around throughout the breeding season. It was initially as spooky as most of the other cats have been. However, we began to notice that it was actually getting along with the lambs that were being kept in the barn during breeding. Of the entire flock, only the lambs had never experienced previous "cat panics". The cat was also a bit calmer than its predecessors. Throughout October and into November, the cat and the lambs became good friends. At that point, the cat also decided that the shepherds were also adoptable. We were faced with the end of the breeding season and a friendly, calm cat in the barn. When the adult ewes returned to the barn, the expected sheep panics occurred when the cat made its appearance. However, the cat did not routinely explode and run wildly from the barn. In addition, some of its lamb buddies continued to stick around it. Gradually, over the first month after breeding, the adult sheep adapted to the cat and calmed down. The mouse population plummeted and the farm now has one more welcome and hopefully permanent resident. It has been a pleasure to watch the cat and sheep adapt to each other. It has been difficult for some of the older sheep to jettison their old, negative memories of previous cats. Nevertheless, they are seemingly willing to add the current good experience to their accumulated memories. So too we have been enriched by our new additions to the family that is our farm. We look forward to the interaction between cat, chickens, sheep and the lambs due to arrive in a little over two months...something more to add to all our memories.

Lambing Time Again

Lambing at Whitefish Bay Farm is just beginning as I attempt to write this piece. By the time I finish writing, I know that we will have our first lambs on the ground. I therefore apologize ahead of time in case there are a few iodine stains on the paper or if it is a bit soggy from birth fluids. I suspect that a good number of readers are also in the throws of lambing or kidding at this time of year. I would also guess that many are much more experienced in the process and are far more knowledgeable than I. To many, what follows will not seem new or dramatic. I often wonder how many readers have been through a lambing and what range of experiences their years have covered. I often wish that I could hear more of such histories and anecdotes.

For us, this is our sixteenth consecutive lambing. (I do not include in that number my younger days as a 4-H Club "sheep raiser". During those three or so years, I never witnessed any of my ewes deliver a lamb. It just "happened". The lamb arrived overnight and was well cared for by the time I got there for morning chores. It is a testimony to the hardiness of sheep that any of those lambs survived my early efforts!) Over the last sixteen years, we have learned a tremendous amount from our experiences. Hopefully, our ewes and lambs have benefited from our increased knowledge.

So much of our lambing experiences are based, in part, upon our choices of geographic location and time of season. I should start by describing the basic conditions under which we have chosen to lamb, since these conditions are often different from those in various parts of the world and different from those in the upper Midwest when lambing occurs earlier or later than ours. We have chosen to lamb in March. When we first started raising sheep, our lambing dates started in very early February and often ran into late April. In Wisconsin, February and early March can still be the dead of winter, meaning that we can experience occasional periods of sub-zero temperatures and deep snows with occasional blizzards. By the time we reach late March and early April, temperatures may try to climb into the 50°F range and there will be strong hints of spring with grass breaking dormancy and trees budding out. Under such climate extremes, it is not logical to lamb unless one can provide good shelter and at least guarantee some degree of warmth for the ewes and lambs. Such requirements also apply to us as shepherds when one is faced with the prospect of limited sleep and long hours with the sheep. We would not have chosen to breed the ewes for this lambing time if we did not have adequate housing for them in late pregnancy and into lambing. To accomplish a healthy lambing under these climate conditions, therefore, requires good housing.

Our sheep housing consists of a 100 year old German dairy barn, a timber frame building whose lower level is cut into the east face of a hillside. The style is often referred to as a "bank barn". We have modified what originally was the milking area (the lower level) into our lambing quarters. The hay mow is in the floor above, which from the west is also at ground level, making the unloading of hay bales easier. The full hay mow thus provides insulation, as does the western wall as it is dug into the hillside. The biggest health danger for such a barn is making it too airtight and thus limiting the exchange of fresh air. There are plenty of windows and doors on the east and south sides and openings on the west up into the haymow providing substantial air flow.

With the experiences of a number of years lambing within these climate conditions and with our given housing standards, we gradually have fine tuned our lambing dates. Our first major modification was to alter breeding dates so that, under normal circumstances, we would not begin lambing until the second week of March. We found that lambing in late February was consistently too cold for the ewes, lambs and us. This change has not totally eliminated extreme cold weather lambing, but it has moderated it. By moving the earliest lambing date to about March 7th, we have also found that we can begin grazing the ewes and lambs in early May (if we have a "normal" green up). With lambs born earlier, we were putting too much good quality baled hay into their growth before they could start grazing.

Our next fine tuning of lambing dates was targeted toward the end of lambing. For our first years, we were leaving our ewes in with the rams for at least 3+ heat cycles. We did this because not all the ewes were getting bred in the first and second cycles. This policy assured that most of the ewes were successfully bred, but it also assured us a prolonged lambing time. The lengthy lambing had three major negatives. First (and perhaps foremost) was the fatigue the two of us felt by the end of lambing. Secondly, the lambs' age range of two or more months tended to compromise the growth of the younger lambs; they faced too much competition from the older lambs. And lastly, it meant that we were still lambing just days before we hoped to start grazing. The solution was simple: just shorten breeding time. It meant, however, that we had a group of ewes that were not getting bred. We learned to cull. Over just a few years, we have managed to select ewes that get bred within the first heat cycle. Those ewes tend to produce lambs that also get bred quickly. We have now reached the point where we can breed just as many ewes as we did in the past, but are able to have them all lamb in less than a month. It means more effort is devoted to lambing in a shorter time frame. We have reached the comfortable capacity of our barn. We know that we can handle (barely) about 90 to 95 pregnant ewes who will lamb between about March 7th and April 7th. If we wish to lamb more ewes, we must either get a larger barn or lamb outdoors at a later date. The two of us are now set enough in our ways (not to mention aging rapidly) that we are opting to stay where we are (or even cut back a bit).

We have sufficient space to set up as many as 16 jugs (maternity pens) along the sides of the barn. We have found that for our larger Corriedale ewes with twins or triplets that a jug needs to be 5' by 7' (with a hanging feeder taking up some of that space. The remaining lambing barn area is devoted to the pregnant ewes until they lamb. As the jugs become full, we will subdivide the main area to create up to five mixing pens for groups of ewes with lambs of similar age. As the mixing pens grow, the number of pregnant ewes is declining. Overtime the mixing pens are gradually combined. Hopefully the overall space balances itself off fairly evenly.

Our normal daily routine, once lambing has begun, is actually not "routine", because it is guaranteed to be disrupted by a lamb being delivered at the least opportune time. If all has gone well the night before, we will get up at 6AM and immediately go to the barn for a new lamb check. Our efforts in this process are aided by a "baby monitor" that we install in the barn each year. It can usually give us a good audio hint up at the house as to what might be going on in the barn. After caring for any new births that we discover (getting the ewe and lambs into a jug and performing the usual "clip, dip, strip and sip" routine), we will slip back to the house for a cup of coffee before the rest of the flock is awake. We will then return to the barn to feed any bottle babies and then feed the rest of the flock. If nothing else is occurring in the barn, we will take care of other barn chores (bedding pens, moving ewes and lambs from jugs to mixing pens, etc.) and then spend the rest of the morning catching up on the non-sheep chores of the day. Bottles are again fed at a noon time barn check. After noon, we will weigh, tag, dock and castrate any lambs that are over 24 hours old. We will also weigh any other lambs that are at certain ages or when we are concerned about growth or health. We keep records of birth weight and at 4, 8, 15, 30, 60 and 90 days. After the afternoon lamb care, we will jacket any ewe whose lamb is over 48 hours old so that the lambs can become accustomed to their mother in a jacket before they move into a larger mixing group. (The mom has been without a jacket from shearing until lambing, a brief period in which the jacket would be counterproductive.) In a perfect world, we like to keep the ewes and their lambs in a jug by themselves for three days before putting them into larger mixing pens. This time period gives the family group a chance to bond and for the lambs to become strong enough to prosper in a larger world. For yearling ewes with their first time lambs, ewes with triplets or other ewes with lambs who have somehow been compromised, we try to extend the jug stay a day or two. From here the family goes to a mixing pen with eventually 16 to 18 ewes and their lambs. After perhaps 5 days together, the mixing pen will merge with another such pen, until eventually everyone who has lambed is together in one large group, hopefully in time for the spring turnout onto pasture in another week.

Evening chores begin at around 5^{PM} with bottles first, followed by grain and hay for the ewes and the rams. A bedtime check at 10 ^{PM} includes the last bottles of the day. If nothing seems imminent at 10 ^{PM}, one of us (usually "lucky" me) gets up to check the barn for pending deliveries at 2^{AM}. And so it goes, until we are over the bulk of lambing. Of course the above schedule tends to get mixed up considerably if a couple of lambings occur simultaneously, especially in the middle of chores, or if we have two or three ewes who decide to go into labor sequentially between 2^{AM} and 6^{AM}.

Our ultimate reward

This routine may sound organized on paper. In fact, it ends up seeming like a few weeks of organized chaos. Mix in a steady stream of short periods of sleep and we start to feel quite "punchy" after the first full week. There are times when it would be so easy to just go to bed and sleep until you woke up on your own. But there are rewards to the chaos. The successful, unassisted delivery of a set of twins, who are quickly up on their own and learning to nurse from a devoted mother is sufficient payback. The excitement of an unexpected color pattern showing up in a birth is a bonus. Hanging out with a couple of old retired ewes while we wait out a birth is relaxing therapy for us and the old girls. If the night sky is clear at the 2^{AM} check, we may be greeted with a beautiful starlit sky or perhaps a show of northern lights. If the weather begins to warm as lambing progresses, the night quiet may be broken by the songs of spring peepers in the surrounding wetlands. At sunrise, we anticipate the first arrival of the migrating Tundra Swans, trumpeting their arrival from the southeast over Lake Michigan at first light. At midday, the first Eastern Bluebirds to arrive for the season also warm the heart. There are definitely pay backs to the job!

I hope the preceding account does not sound too disjointed. I began writing just as the first lambs arrived. Now after 15 days (I think), I am hoping to finish the narrative. It has been busier and more intense than normal so far. To date over 100 lambs have arrived and we are just past the two thirds mark in terms of pregnant ewes. It looks like another ten days and we will be back to normal.

Fleeces For Sale

In all likelihood just about every person reading the Black Sheep Newsletter is somehow involved in fiber production by either sheep, goats, camelids, bunnies or other species. As such, we may raise and care for the fiber animal; use their fiber ourselves; sell or give it away; exchange or buy the fibers from someone else. While we have such common shared interests, it is also interesting how the manner of exchange of the fibers has changed over a relatively short space of time. There are so many variables in this complex equation of fiber production and selling that it is very difficult to make any generalizations. Recently, I have been thinking of our personal experiences and how they have changed since our first fleeces were shorn in early 1991. We are by no means veterans in this activity, but we have now been around long enough for me to make some observations about the process and how it has changed from our perspective.

From a production standpoint, our farm has evolved greatly. We began with a flock of twenty ewe lambs and a single ram. Their fleece quality varied greatly and our care for the sheep and their fleeces has certainly matured since that time. Our first shearing turned out to be a nightmare. Our first shearer came well recommended, but it soon became apparent that he usually worked with meat breeds and had no concern for a quality shearing job. I believe his motto might have been something like: "Shear 'um quick; who cares about second cuts and bleeding sheep". Even knowing as little as we did at that time, we quickly realized that our first clip was a disaster in terms of quality. We spent days skirting 21 fleeces, trying to salvage something that looked marketable. This disaster was impetus for me to learn the art of shearing. Over the next three years I sheared our slowly growing flock, with more concern than our first shearer, but certainly without his speed. The end product improved each year as my skills progressed. Gradually, we had a better product to offer. It also became apparent that I was not getting enough shearing experience each year to improve as much as I had hoped. We made the decision to find a quality shearer and to put my aging back into semi-retirement. We were fortunate enough to make contact with David, a younger man than I, and a fulltime shearer who also had a great sense for a quality fleece and the work needed to produce it. We have been lucky to have him shear for us ever since. This change has allowed us to devote our attention to fleece improvement in other aspects of our operation.

Providing good support for Dave's shearing was the first step. No matter how good the shearing job, the freshly shorn fleece can still be ruined by poor handling. We now have a "professional" shearing crew (usually five or six in number). Our efforts involve: 1) catching each sheep and removing its jacket without contaminating the fleece, 2) keeping the shearing area swept constantly, including the immediate bagging of all loose trimmings, bellies and leg wool, 3) doing an extensive first skirting as soon as the fleeces is off the sheep, and 4) bagging and identifying each fleece as soon as it is skirted. The shearing of about 130 sheep each year takes just under two days. It is a time of hard work and good companionship and camaraderie.

The quality of our fleeces has improved with time and experience. We now do a better job of jacketing our sheep and keeping the jackets in good repair. Our pastures have improved in quality with a resulting decrease in plants that produce burrs and stickers that would otherwise compromise the fleece. Our breeding selection has improved the overall quality of the fleeces. As a result, our product has not remained static; a significant factor when it comes to selling the fleeces.

Even if our product had remained static, the last sixteen years have seen a dramatic change in how we sell our fleeces. When we first tried to market our wool, our options were few. We could try approaching our local spinning guild, but their membership was small at the time and they already had their "preferred" sources of fiber. There was always print advertising in, for example, the spinning and fiber publications. Even then classified ads were relatively expensive, especially for new producers like ourselves who still had just a few fleeces to offer for sale. We became involved in a co-operative fiber group. By pooling our resources we managed to produce a brochure that listed each participant and their product. By sharing the cost, it became affordable to advertise the co-op brochure in the fiber publications. And so, this was how we began. It was a start but a very slow one. Over the course of an entire year, we rarely approached selling all of our fleeces, even as they improved.

Another early option for sales was attending fiber shows and sales, where vendor space could be rented. Our farm's relatively remote location, along with the nature of our other activities, prevented us from being able to travel too far or for any length of time. In turn, that limited our participation in shows, both as competitor and/or vendor. Eventually, such a fiber event was organized within a three hour drive from the farm. In our third year of shearing we had sufficient number of fleeces and experience that we felt we could take a chance on renting vendor space. We loaded up the back of the pick-up truck the night before with as many fleeces as it could hold. The next morning I left home early for a new adventure. The event was within easy driving distance of Chicago, northern Illinois and southeast Wisconsin and drew a large turn out of spinners and fiber enthusiasts. It proved to be our first big sales success. It meant a late evening drive back to the farm to unexpectedly load up the rest of the fleeces so that I could return the next day with sufficient inventory. From this event, we began to develop a customer base for future mailings and a newsletter. It was indeed exciting for the two of us. We believed that we were on our way and we had great plans for the following year's show. What we did not count on was that the show was cancelled the following year, never to reappear!

Partly in response to the cancellation and to the general lack of other fiber outlets in the area, a small number of fiber producers from our county came together in the summer to form a loosely knit organization. One of our primary goals was to organize our own annual fiber market. The following spring the group held its first ever Door County Shepherds' Market. It began as a single day event, whose sole purpose was the sale of fiber and related products that were raised within our county. The event was a success from its inception. Sales were good for most members (at least those who had a quality product to which they were committed). The event has now continued and grown over 12 years. The group has grown and evolved and the event now runs for two days. For each vendor successes have varied, and for many of us it had become our primary outlet for our fiber products. For the two of us, it marked the first time that we sold most of our fleeces within two to three months of shearing. Our customer base also grew such that many of our buyers from the Market would reserve the same fleeces the following season.

One of the especially nice aspects of selling at such a market or fiber show is that, as sellers, we get to meet our customers in person. The customer also gets to select their fleece purchases by being able to see, touch, feel and handle the wool. The uncertainty of mailing samples is avoided. The chance for a dissatisfied customer is limited. The cost and aggravation of shipping bulky fleeces is negated. Getting paid is not a problem. The biggest "difficulty" is geographic. People, generally, will only travel so far for such an event. (Over the twelve years of the Market we have, however, found that customers are coming greater distances each year, perhaps due to the permanence of the event and the good reputation that it has developed.) From a personal standpoint, the Shepherds' Market became our primary outlet for selling fleeces. There seemed little more that we could ask for marketing wise.

Despite the success we experienced from the Shepherds' Market, a curious development began simultaneously. It would eventually change just about every aspect of our marketing strategy. The development was the maturation of the Internet as a method of communication. We got on the Internet bandwagon relatively early, especially for being in a rural area. Our primary intent was to use a website to publicize our other businesses, a bed and breakfast and an art gallery. We have had a functioning website since 1996. Publicizing our sheep and wool operation was, at that time, only a secondary consideration. Nonetheless, we began offering fleeces for sale through our website as of 1997. Our sale of fleeces through that media began slowly and then progressed rather rapidly. It has now reached the point where virtually all of our fleeces are sold online. Usually we sell almost all the fleeces within 48 hours of posting them to the website, usually sometime in April each year. In many ways, it has been a blessing for us, business wise. It has also produced its own unique set of headaches. The first headache was that we no longer had sufficient number of fleeces to sell at the annual Shepherds' Market. We have had to re-tool our approach for our portion of the Market, to emphasize 1) the yarn and roving that we have made from our "good" skirtings, and 2) the finished products of our labor, i.e. handwoven and handspun items. As the host farm for the event, we can at least turn a little more attention to the overall running of the Market.

The second headache of Internet sales is its relative impersonality. For both the seller and buyer of wool for handspinning, the process of selling/buying a fleece needs to be very tactile. No matter how much you try to describe a fleece in words or with photos, it never quite conveys it complete character, in terms of color, luster, softness, handle and all the other intangibles related to the "spinablility" of the individual sheep's fleece. It is a leap of faith for the buyer to purchase a fleece without so much as receiving a sample. Due to the success of our online marketing, it is no longer realistic to send a prospective buyer a sample of a fleece. By the time it is mailed, received and analyzed by the prospective buyer, the fleece will likely have been purchased by someone else. Under these circumstances, one needs to stand behind the fleeces and be willing to accept them back if they are not what the buyer wanted. Luckily we have had <u>very few</u> returns.

Also lost with internet sales is personal, face to face contact. We have always found such contact to be one of the rewarding aspects of fleece sales. We have, nevertheless, made some lasting friendships over the years based strictly on Internet contacts. Yet, we always look forward to the day that we might meet each of these customers. Often lost with the personal contact is a degree of personal responsibility. It seems that every year someone (old customer or new) decides, for whatever reason, to not finish a sale. The fleece is set aside awaiting payment. Eventually it becomes apparent that the buyer is not buying, but there is rarely an email either apologizing for or explaining the change of heart. It would be nice to receive a note saying that they have reconsidered. This sudden, mid-sale silence is frustrating and discourteous, but it seems to be part of the Internet "morality". Inevitably, it leaves us with a bad taste for the individual and a fleece that is now unsold. The unsold fleece often goes back on the website as a "one and only" listing and becomes difficult to sell. After all, something "must be wrong with it".

In some aspects the "old ways" are better. Perhaps that is why many of us have gotten into a life style that should be a bit calmer and more in tune to the seasonal rhythms. The old ways may have been slower and they did have their own frustrations. They did, however, have more of a personal touch to them that is now so often lacking. I am sure that the market for natural fiber for spinning will continue to evolve, probably in ways I cannot imagine. In the meantime, we will continue to raise the sheep, enjoy their company and marvel at their beautiful fleeces.

Thoughts on Retirement

As the summer progressed, it became clearer that Oscar would become all the more difficult to care for. He was having difficulty performing the work he had previously done so well for us over the years. By the end of the summer, it was obvious that we could no longer provide the care he needed, without the help of specialists. Such help would be expensive and would come with no guarantee that Oscar would not start showing other problems associated with advancing age. In late August, we made the decision to sell Oscar to someone who was better suited to provide the kind of care he needed. In return, we would purchase a new, younger and (hopefully) healthier replacement. Phone calls were made; papers were signed; a replacement was located. On a cold, gray September day, Oscar left the farm for good.

Oscar

Oscar had been with us since we began farming. He came with the farm when we purchased it in 1983. How long he had been here prior to that I do not know for sure, but it was probably sometime in the early 1970's. Based upon his registration number, he was born in the early 1960's. As a reader you were first introduced to Oscar (but not by name) a few years ago when I related our experiences with a ram lamb, Nicely Nicely. When he was about six months old, Nicely Nicely and his fellow ram lamb buddies had the opportunity to inspect a tractor for the first time. As teenage farm boys with time on their hands are want to do, they seemed to enjoy discussing tractors, an activity often associated with kicking tires, or in this care, butting tires. Nicely found tire butting especially enjoyable and he was especially fond of the tires on our Massey Ferguson tractor. That tractor had long been known in these parts simply as "Oscar".

Buy today's standards, Oscar was a relatively small tractor. Yet, in his prime, he was the main source of power on a farm that not too much earlier had relied on true horse power. He pulled a three bottom plow ever year, along with various other cultivation discs and harrows. He pulled the drills for planting. He powered the haybine and haybaler for the cutting and baling of tens of thousands of small bales of hay needed to feed the dairy herd. He was the quintessential tractor of the small Midwest farm of the 1960's and 1970's. I know that, if he could have spoken, he would have been able to talk of being present for most of the joys and tragedies of the farm family that preceded us here. Oscar enjoyed a reduced work schedule when we became stewards of the farm. We do not grow our own grain, so Oscar spent most of his time with us caring for our cherry orchard and cutting, raking and baling our hay. He already showed signs of a rugged, hard working life when we came to the farm. Yet I remember starting him for the first time, after he had sat idle for well over a year. I did not expect his engine to turn over, let alone start. So, I was shocked when he started as soon as I turned the key. It was the first time I had ever sat on, let alone driven, a diesel tractor. At the time, the experience took my breath away!

I came with minimal/nonexistent diesel maintenance skills, so we relied on the local Massey Ferguson dealer to perform any major repairs. Even with the good quality care we tried to provide, this year it was becoming evident that he would soon need a major overhaul and/or rebuilding. It would have been an expensive proposition, perhaps even more so once the disassembly began when other problems would probably be found. We opted to transfer his care and rehabilitation to Dean and his boys at the implement shop. That transfer, plus an additional outlay of cash, brought a much newer and healthier tractor to the farm this fall. We have yet to christen the new unit with "Oscar II" or any other name. So far, it has not developed a personality. It is our hope that the replacement tractor will also be the last tractor we ever need to purchase. If Oscar II lives as long as his name sake, he should easily out live me!

Why, you might ask, have I spent so much time here writing sentimentally about an old tractor in a publication devoted to sheep, goats and fiber production? For Gretchen and me, this tractor has had a more significant impact on our operation than any other single object, be it animate or inanimate. Without Oscar, we would not have been able to bale hay. If we had to rely on purchased hay, our operation would have been vastly different. On a personal level, I am amazed to think of the hours I have spent working atop Oscar. It ranged from the time of long, hot dusty days to hours in snow. Yet it was also time devoted to both serious and trivial thought. Whether cutting, raking and baling over 3000 bales a year or spreading the composted manure that results from that hay, it all leaves lots of time for thought, all of which was spent with the steady rumble and roar unique to that tractor. I will miss Oscar in his retirement. It will be a long, long while before I feel as comfortable with his replacement. Ironically, I spent my last hours

200

with Oscar, finishing up second crop hay and musing about retirement of a different sort.

Old and New Sheep

At virtually the same time that we were having to make decisions regarding Oscar's fate, we were suddenly faced with another major, unrelated decision affecting the ongoing operation of our farm. Since lambing, we had been toying with the idea of downsizing our flock. Like Oscar, the two of us were feeling the effects of many years of farming. The arrival of 145 lambs in a period of just over three weeks this spring confirmed to both of us that we perhaps needed to scale back our efforts a bit next spring. Eventually, after the rush of lambing had passed, we made plans to not keep as many ewe lambs as we normally would as replacements. We might even sell a few of the adult ewes who, for one reason or another, were not at the top of our "favorites" list. In addition, the number of ewes we would breed in fall would be drastically curtailed. Next spring was already beginning to sound like a virtual vacation. But, as usually occurs, after we had recovered from the burnout of lambing and progressed into the heart of summer, the drastic cutbacks began to fall at the wayside. There were just "too many" nice ewe lambs. We "certainly" could not part with a number of them for all sorts of "good" reasons: that one looked like she would have a wonderful fleece; this one was the last of a family line; and (the most scientific of all) she was just too cute! Are we the only shepherds who suffer from this disease? By the time the last of the lambs that we did choose to sell had left the farm, we still had more lambs than we had planned on after lambing.

Next followed the corollary rule for adult sheep: you always fail to cull as many adults as planned. The decision to sell an adult can be especially difficult. The ewe who still refused to be a good mother after getting a second chance easily seals her fate. What about the older ewe, who met our retirement age and who, if healthy, would get to live out her life with us if she was friendly and easy to work with, but who steadfastly refused to be friendly? Does she go or does she get to stay because: 1) she has a nice fleece, or 2) she gave us lots of nice lambs? When all those decisions were made, we had progressed a little more toward our reduction goal but not as drastically as we once had talked. The previous year we had entered breeding with 132 adults. This year the leaner ledger showed 120 adults with a larger percentage in our retirement group. Hence, we had made progress, albeit not as drastic as originally envisioned, yet progress nonetheless. To salve our consciences, we also decided not to breed the ewe lamb population for the first time, thus further reducing the number of pregnant ewes next spring. Doing so would, hopefully, eliminate some of the more labor intensive, first time lambers. Given another year to mature does seem to lessen the problems the first time mother can experience.

We were not about to retire ourselves, but perhaps life would be a little more leisurely next spring. By the time we had come to the decision to replace Oscar, it was August and the prospects were brighter for at least a slightly smaller flock and an easier lambing. We had even planned some time off. I was ready to travel to Denmark for over a week to visit adoptive family and friends and to check out a number of flocks of sheep and goats. Later, Gretchen would head for California for the Spin-Off Autumn Retreat. However, another retirement got in the way of just about all of these plans except for Oscar's departure.

Out of the blue, in August, we received a phone call from Martha, a dear, long time friend and fellow colored Corriedale shepherd. Many of our original colored ewes came from her flock. She had recently made the very difficult decision to retire from raising sheep to concentrate on other loves. She was calling to offer us virtually all of her ewes. It was a decision that we needed to discuss rationally. We promised to spend the afternoon discussing the situation so that we could get back to her quickly. In some ways, it was an easy decision and in one way a very difficult one. We were familiar with the flock and the way in which they were managed, which was exemplar. Her flock had been closed nearly as much as ours, so that our health concerns were minimal. We suspected that even without looking at them that, as a group, these ewes would not only complement our own flock but also add to our quality and diversity. The difficulty? If we obtained her flock, or even a portion, our plans for reducing our flock size would fly out the window. So much for planning; we called back in an hour to tell her we would drive across the state the next day to have a look at the bunch. The next day we did chores early, hung out our "Closed" sign, and were on the road to southern Wisconsin to visit with old friends, look at the individual members of her flock, make our decision on which ones to purchase and still be home before dark to bring in the sheep and to do evening chores. When all the dust had settled, we had committed ourselves to purchase nearly all of the ewes who were still young enough and healthy enough to be bred. As part of the agreement, we would purchase all of these ewes' lambs, as they had not yet been weaned. In two weeks, we would return with our stock trailer to pick up 36 sheep. In one sense, we were lucky that Martha had already, over the last few years, significantly reduced the size of her breeding flock. Had she not done so we would not have been as much help and we would have been faced with more difficult decisions.

Deciding to buy the breeding flock was, in reality, an easy one. Besides our confidence in their health and quality, it just seemed to be the "right thing" to do. We knew that Martha in her own way had put at least as much toil, care and love into her flock as we hope we give to ours. To somehow break up that group was not right; worse yet was the prospect that some of them might have had to be sent to market. To make a decision to turn away from so many years of a strong commitment is difficult.

As I now write, it is nearly December. The new ewes have been with us since September. Their lambs have been weaned and either sold or added to our flock. The addition of this group of ewes and lambs to our flock went surprisingly smoothly. The new ewes were "accepted" by our flock when we eventually combined the two groups. The daily routines for the two flocks were (as expected) different. It took a while for the new contingent to adjust to our daily routine, especially the task of heading out to a new swath of pasture each day. At first, they were obviously puzzled as to where they were going and as a result they always ended up at the end of the daily parades. They were not habituated to our portable electric fencing and it took a time before the ewes and especially the lambs learned that the strange strands carried a strong shock. For over a month, the new group tended to bed down as a separate group in the barn. It was disconcerting for the two of us to hear sheep voices that were somehow "different" than the ones to which we were accustomed. Now, after three months, I am no longer aware of the different "baahs" and "maahs".

We put our rams in with the ewes in mid October. The new ewes were split into three subgroups and joined some of our ewes in three separate ram groups. The remainder of our ewes were split between the other four rams. It was still possible to witness different behavior patterns, even within these breeding groups. There was also a lot of cross-pasture communication between our imported ewes. Obviously, even then, the new ewes did not feel as if they were totally part of the larger flock. One thing was however certain: the rams did not care where the ewes came from! They bred and marked the "new" ewes just as eagerly as the "old" ewes. This year was intentionally to be the shortest breeding period we had ever given the rams. It was almost as if the rams knew that they had exactly two heat cycles with the ewes and no more. They seemed to make the most of it.

As we enter winter, where do we stand as to our "retiring" plans? The flock is now one sheep larger than it was at this time last year. We stuck to our plan to not breed the lambs, so the pool of potentially bred ewes is a little smaller than last year. However, since all of the ewes that we bred were adults matched to proven adult rams, the overall conception rates will probably be up. Despite shortening up the breeding period, few of the ewes were re-marked so there is a good chance for nearly all the adults being pregnant. We could consider ultrasounding the flock in a couple of weeks. It might help us with their rations, as we could separate the open ewes from the rest. However, we will probably just wait for shearing when we can make a visual assessment of who is carrying lambs. No matter how we approach it, we will have a busy three weeks in March.

We are glad that we made the decision to purchase the ewes. The trip to Denmark was the major fatality of the twin purchases of tractor and ewes, but it can be rescheduled. Gretchen still worked in her week of spinning out West. Of more importance, it is our hope that we eased someone else's decision to distance themselves from raising and caring for sheep. I know that it was very hard for Martha, placing the last of her animals on a trailer to leave her farm. It means a different emphasis will be placed in her life's activities after many, many years with sheep being an integral portion of it. We are comforted by the fact that we know that she looked forward to the change, no matter difficult it was.

All this does make me ponder what a similar retirement circumstance would be like for the two of us, when and if it does occur. The sheep have become such an integral part of our existence for such a long time that it is extremely difficult to picture life without them. It gives us reason to think even harder about the circumstances, should we have to, by choice or necessity, give up our flock. It would be my hope that we will be in the position to be able to find a good home or homes for all of our flock should that time come. Anything less will make the decision to "retire" extremely difficult.

Choosing a Breed

Not long ago we had a visit from a young couple who wished to discuss their possible future in sheep raising. For me, it is always interesting to hear a new perspective, especially when so much of it is surrounded by the dreams of youth. This particular pair, while it was obvious that they had stars in their eyes, had at least already done lots of research and planning. To their credit, they had already sought out expert advice and had plans to further pursue the ideas which they had formulated. Picking our brains may always be a questionable practice. Nonetheless, they came to us, primarily to gain our perspective as graziers (with an emphasis on sheep). Their plans centered on starting a flock of milking sheep, using a management intensive grazing system, with the ultimate goal of making and selling their own artisan cheese. They also had an interest in fiber production, but it was currently secondary to dairying. It was a pleasant way to spend a few hours on a late winter afternoon.

Over the years we have had similar visits. In most cases, the center of interest relates to wool production. Such visits range from folks with little concept of what they might wish to do, to others (like this most recent couple) who had already researched and learned a good amount. Often the key question asked is why we chose Corriedales as a breed of sheep. I am sure that is the same question regarding which breed of sheep to obtain that is asked of many shepherds. Rather than try to write on the pros and cons of raising Corriedales, I thought I would approach the question in a different fashion. I will attempt to describe how we arrived at our decision to raise Corriedales. In a similar fashion, perhaps others can formulate their own plans regarding a breed (or breeds) of choice when beginning a flock. Bear in mind that what follows is only our line of thinking; I am sure that every established shepherd will have their own unique perspective.

When we decided to raise Corriedale sheep, there were two separate areas which guided our choice. The first area I will term "breed specific"; the second relates more to "personal circumstances". Also bear in mind that when I write of Corriedale sheep I refer to purebred Corriedales but not necessarily registered Corriedales. This differentiation is due to the fact that we raise both white and naturally colored Corriedales. The presence of the traits of color genetics precludes a purebred Corriedale from being registered as a Corriedale in the U.S. through the American Corriedale Association.

Breed Characteristics and Concerns

Prior to a final decision as to which breed we might pursue, we had to think about possible alternatives to a single breed. There seemed to be two ways for us to proceed. We could concentrate on a single breed and work strictly within its purebred boundaries. Or, we could look into starting with more than one breed and concentrate on either a fleece "type" or "style" and/or seek to produce a variety of fleeces using different breeds and cross-breedings. We opted for a concentration within one breed. It was a decision made in our early, formative days of sheep. We felt better about becoming knowledgeable about and comfortable with one breed, rather than initially confusing our "education" with the variety that multiple breeds would offer.

When we sought out a particular breed, it was with the expressed desire to raise sheep primarily for fiber production. Specifically, we were interested in a breed that produced wool that would find a demand in the handspinning market; we sought to use the wool for our own spinning and weaving and also to sell to people of similar interests. We found that the Corriedale breed had an established record for its relative ease of spinning. The wool is moderately fine, soft and has a good crimp. The final yarn product is fine and soft enough to make it comfortable to wear next to the skin. When the breed was first developed in New Zealand and Australia, one of the original goals was to create a sheep that would produce the wool of a Merino on the meat carcass of a Lincoln. In a sense, it was an effort to produce Merino quality wool on a sheep that could also be successfully marketed for its meat (the traditional Merino's shortcoming). The result was not perfect: the fleece was not quite as fine as a Merino (but still relatively fine), but there was a significant improvement in meat production. Over the years, and especially of late, the commercial emphasis in sheep has shifted dramatically toward meat production, with less emphasis on wool. As a result the Corriedale fleece has gradually become less fine than it used to be. Currently the micron range for Corriedales is from 26 to 33. It is still, however, possible to breed for the earlier fineness if that is one's concern. We have been able to lower our average micron range into the low to mid 20's. (It was exciting for us to see Corriedales in New Zealand in 2004. It was further exciting to see, and feel the national champion ram fleece from a fine animal from the Coldstream stud, that tended to buck the current trend in New

Zealand toward a stronger wooled Corriedale. Its hogget fleece tested at just over 22!)

We also sought a breed that individually produced a significant volume of wool. Again the Corriedale breed met that need with adult ewe fleeces weighing from 10 to 13 pounds. It would do us little good to produce small amounts of wool unless we could add a large price differential to it. Ultimately the wool had to be marketable. Our research, at the time we began, indicated that there was a market within the handspinning community. It has remained so today. From a marketing standpoint, we also hoped to avoid some of the "boom and bust" breeds. Over the last 17 years, we have seen a number of breeds (usually initially rare) become exceedingly popular, with fleeces and breeding stock first demanding fantastically high prices. Inevitably, in a few years that popular breed would be replaced by another "rare" breed and the cycle repeated itself, while the previous "breed of choice" experienced a glut and falling prices. It must be an exciting roller coaster ride, but we wished to remain more sedentary yet secure.

Our other fleece concern related to "jacketability". We planned on jacketing our flock to retain the clean quality of the wool and to reduce sun-bleaching of the colored fleeces (and thereby hopefully increasing the wool's value and marketability). Not all breeds do equally well under jackets. With some breeds, the fleece will tend to either pill, felt or unduly wear if subjected to a jacket year round. Corriedale fleeces tend to behave very well under a jacket if the jacket is properly fitted and in good repair.

Lastly, as far as wool was concerned, we sought a sheep which had a history of being able to produce colored offspring. In the United States, the recessive colored genetics which occasionally express themselves in white Corriedales, have not been eliminated. For us that is a plus, as it meant that we could pursue both white and colored sheep of the same breed. It also means that a breeder of white Corriedale may have a colored lamb show up on occasion. In many of those cases, the lamb is not seen as desirable to that breeder, but it can often be an addition to the gene pool of a flock such as ours.

On the non-wool side, we sought a breed that was suited to grazing. The Corriedale is a hardy animal that is adapted to a wide range of conditions and environments. Because we planned on using a rotational grazing pasture system which required the flock to move to a new paddock each day of the grazing season, we also needed a breed which tended to have a strong flocking sense, rather than a breed which tended to disperse individually at the first opportunity. Corriedales are easy handlers, calm and not too spooky (for a sheep at least!). It is a testimony to their flocking instinct that we can move the flock daily, without the assistance of a herding dog.

Our sheep also had to have a non-wool market value. Very early we decided that we did not wish to get heavily involved with the show circuit and/or the breeding stock market. Therefore we needed a breed which could be sold competitively in the commercial meat market. The smaller framed breeds can do well in some of the specialty/ethnic markets. We do not live anywhere near those markets nor do we live in an area with a large enough population to make direct marketing a good option. So, if we are to sell our lambs commercially, they must have the ability to meet the demands of the commercial market. It is a strange market in some ways, as it is seemingly not consumer driven. Nonetheless to maximize the saleability of our lambs commercially, they need to have the ability to approach a slaughter weight of 120 pounds. With proper management, good pasture (and often a little luck), we are able to coax that type of gain from our Corriedale lambs most of whom tend to be twins. From our standpoint, it is an unfortunate reality, but the commercial sale of our lambs still represents about half of our gross income from the sheep, even with our fleeces selling on average for well over $10 per pound.

Non-breed/Personal Concerns

Many of the other decisions we made regarding choosing Corriedales were not directly related to the breed as such. Availability was a key component. We had to be able to obtain good quality breeding stock in both white and naturally colored lines without having to travel thousands of miles. This concern was partially a financial one and partially due to our desire to be within a "community" of Corriedale breeders. The later was based on our need to be able to share expertise and possibly obtain future breeding stock. In the late 1980's and early 1990's Wisconsin offered a good selection of both white and colored Corriedale breeders, which thus made our decision much easier. (The same decision today would be much different in Wisconsin. We would have to go much further afield to obtain the same quality breeding stock and would not have been able to cultivate the close relationships that we did earlier.)

Cost was another major concern. As inexperienced shepherds, starting out on a very limited budget, we did not wish to invest large sums on a small number of a rare breed, at least until we knew that we could properly care for them and that they would do well in our environment. We also did not wish to enter into the purchase of a particular breed with it being primarily for investment sake. We did not wish to raise sheep just so that we could then sell them to others who also saw them primarily as investments.

212

Certain criteria were a "non-issue" for us from the beginning. Based upon the plans for our life style, being able to participate in show circuit activity meant little to us. Hence our choice of breeds was not show related. Neither did we aspire to become heavily involved in the sale of breeding stock. Whether or not there were breeding stock buyers available for our excess quality lambs did not concern us. As time has passed, this issue has become even less of a concern.

Final Decision

Ultimately, our final decision to choose Corriedales was based upon a number of often inter-related factors. Some factors were breed-specific; some were based on our personal preferences and circumstances at a given time; some were based perhaps just on serendipitous luck. I can think of a few other breeds that might have worked out as well for us, just as I can think of a number that I am glad that we avoided. Had our plans been different, or had they changed dramatically, our decision would have to be viewed in a different light. Many years ago we toyed with a dairy sheep operation. Had we pursued that idea it would be interesting to now view our decisions regarding Corriedales. We always find it interesting to listen to someone who is making their initial decisions regarding a specific breed of sheep or even a specific species of fiber animal. It is equally interesting to hear of such decisions in hindsight from others who are well established with their flock. What road did you follow to arrive at your current direction? What new roads will you take into the future?

New Ewes in the Lambing Barn

June is nearly over and this year's lambing season is well past us. For us, it was a very typical lambing, with its usual joys and trials, late night visits to the barn and lots of lost sleep. It was our shortest lambing on record; all the pregnant ewes gave birth within a three week window. It was an intense experience, but at least our planning during breeding came out as we hoped. All of our lambs were born during March, so now, as I write, most of the lambs are about three months of age. They have been on pasture for nearly two thirds of that time and have already become seasoned grazers. Within another month most of them will be weaned and sold as feeders to good friends, who farm very near us in the next county. The lambs' experience should continue to be a quality one for them.

It will be good to have most of the lambs moving on, as we once again seem to be stuck in a small pocket of drought. The pastures look and behave more like it was late August rather than June. The removal of over 100 lambs from the farm should ease some of the grazing pressures we will continue to face, at least until we receive significant rain. There have been torrential downpours all around us, all spring and early summer. The contrasting effect of more moisture is truly amazing to see within fifty miles or less in any direction (except perhaps out in the middle of Lake Michigan!).

Sometimes, I wonder if somehow I am personally responsible for the lack of rain. Last winter our rain gauge got left out in the cold. It froze and then developed a leak once it thawed. I was not able to get around to purchasing a new one until just a couple of weeks ago. If I was superstitious, I might believe that the weather gods figured they need not rain on us until we got a new gauge. The new gauge is now in place, so it is permissible to rain on us again!

Half of the first cutting of hay (what there is of it) is baled and stored in the barn. I am taking a break prior to tackling the remaining standing hay, in the optimistic hope that it may rain on us for a couple of days. During my hours on the tractor cutting and baling, and now in rest and recuperation, I have been able to look back on the last few months and make some observations regarding how life with the sheep has been different than in previous years.

We are now nearly ten months into our major flock additions. Aside from the fickleness of the weather and its resulting effect upon us over the years, adding new sheep to the flock has had some of the most major consequences we have experienced, certainly more than we had anticipated. For those who may not know or remember, we purchased 14 ewes and their 26 lambs last September. They are the first sheep to come to the farm outside of our own breeding in the last 12 or so years. We expected that this new sub-group would be different in character and behavior than our own sheep. They were, after all, strangers to our flock and to our flock's way of operation. Superficially, the new ewes have adjusted well, some more than others. We had not anticipated how so much of our routine has been standardized for our flock (rams, ewes and their lambs) over the years, but then it had been 12 years since the last ewe arrived from outside the borders of our farm.

I need only look at our rams (all home grown) to realize how much of a routine our sheep get into after a number of years.

Currently we have seven adult rams, ranging in age from Marvin and Misha at six years of age to Stud Muffin, the youngest at two years. During most of their time on pasture in spring and summer, they have access to a large single pasture which has some good spots for grazing and plenty of shade for relaxing when it grows warm. The pasture is still known as "Number 4" as we have yet to give it a more creative name. Perhaps we should call it the "Rock Garden", since it boasts an unusually dense concentration of large stones and boulders, which I am certain dissuaded the original farmers from ever thinking of clearing and plowing the area. It is not a very direct route for the sheep to get from the barn to Number 4, largely due to my own poor initial planning. It is an "add-on" pasture, fenced later than the rest. It requires walking through two smaller pastures and making a very sharp turn once into the second of those two pastures. (It could have been a "straight shot", but shepherd planning of gate placement was poor at best!) When using the same pasture for our ram lambs, the only way to get them there is to set up a runway of portable electric fencing, down which I must herd them doing my best imitation of a Border Collie. Even with all the extra help, the ram lambs have trouble with the navigation. This spring however, our "old boys" warmed my heart. On the first day they were able to get to Number 4, all I needed to do was to open gates for them and they were set; no detours or stops in the smaller pastures on the way, just a steady yet eager walk with me trying to hurry to catch up. They had not been there since September and now it was nearly June, but they remembered the routine down to its fine points and corners. In the evening, the return home was even easier. The familiarity and routine (in some cases of five to six years) plays such an important roll. I would also suspect that it has helped to cull the trouble makers, who no matter how old, insisted on picking brief fights on the way out each day.

216

Sheep do change and adapt, but just like us humans, it is difficult for them to throw away old, ingrained habits and routines. This behavior is readily apparent in the evening feeding routine. We bring the flock in from pasture near sunset and feed them extra hay in the barn for the evening. Prior to, during and following lambing, the feeding routine also includes grain. After nine months with us, many of the 14 "new" ewes still have not mastered the "proper" technique for getting queued up in the evening at one of the many five sided feeders to get a prime eating spot. It can probably be argued that the local ewes have become so good at it over the generations that it is hard for a "foreigner" to break into the routine.

Similar differences between the 14 "new" ewes and the rest became apparent during lambing. All of our ewes were born in our barn, under the same type of space and weather constraints, year after year. The same applies to their offspring and subsequent offspring. They have become adjusted to the two of us being in and around the barn throughout much of lambing. They have adapted to lambing close to each other and rely on us to separate them, post-lambing, from the rest of the pregnant flock. The possibility of mismothering is great under these constraints. Consciously and unconsciously over the years, we have fixed the mismothering problems through selection and culling, keeping those ewes who do best in our lambing environment. As a result, we have selected for a ewe type that fits our operation (but who might not do nearly as well in a different setting). This year a bunch of new ewes, most of whom were experienced mothers but who were foreign to this environment, were asked to lamb in our barn for the first time. It became apparent that many of these new ewes had difficulty dealing with that situation. Keeping track of multiple lambs right after birth and then later when first in the mixing pens was more of a challenge for some of them than we expected. A number of these ewes with twins ended up with one lamb being "partially forgotten". The problem was then reflected in significantly different growth rates for each of the twins. If the slow grower is a ewe lamb or a ram lamb we decided to keep intact, the selection process away from this problem will be accelerated, since the odds of our retaining that lamb as a breeding replacement are very small.

There will be little room in the flock for replacement ewe lambs this fall. We still plan to get down to a smaller flock, as we had been planning prior to our opportunity to purchase the extra ewes. Thus, attrition and moderate culling still will not create as many places this fall as normally would occur. We do have our eyes on a couple of ewe lambs and at least one ram lamb with dams from our group of 14 new ewes. It will be interesting to see whether they continue to display behavior traits similar to their mothers or if they develop more along the line of our other, home grown lambs.

Postscript to the Tale of Oscar

It may be remembered that along with our purchase of the extra flock last September we also had to retire our old Massey Ferguson tractor, Oscar. A number of you have written about similar attachments. It is fascinating how some of us can become attached to old pieces of equipment that made a critical difference in our ability to successfully farm. In some cases, we may almost feel as though the tractor has its own heart and soul. We certainly get to know it well enough to recognize its personal behavior as if it were alive. Since trading in Oscar for a newer tractor, I have heard from Dean at the implement dealership that Oscar found the kind of intensive "medical" attention that he needed and which I was not qualified to give. A rejuvenated Oscar has already found a new home at another farm. It is a pity that some of our favorite old ewes cannot find similar quality rehabilitation and hence a longer peaceful life. Oscar's replacement has so far provided yeoman service at our farm this spring. I have yet to become accustomed to different controls and quirks than I knew with Oscar. And, as yet, the new tractor has not developed a personality. It has yet to be christened with a name. I am still not sure if Oscar II is appropriate or suitable. But thanks for asking!

A Fresh Perspective – Travels in Denmark

The summer has been a rough one for the flock and for us, as shepherds. In reality it has been difficult for anyone attempting to farm in our little corner of the world. We progressed well through lambing and into spring. Pastures greened up nicely under moderately good spring rains. The ewes and all their lambs were happy to begin grazing in early May, as they usually do. The lambs grew very well on the early pastures; the ewes appeared to be regaining conditioning while still producing a good supply of milk for the lambs. Once grazing began, the rains quickly subsided and then virtually ceased. In June, we recorded one good rain of just over one half inch. For July and most of August, it failed to rain at all. We are dependent upon good rains in the spring and summer, usually recorded in multiple inches rather than tenths of an inch. By the end of June, the flock had completed its second full rotation grazing the pastures; we had harvested our first cutting of hay (about two thirds the normal tonnage). By the first of July, in response to the developing drought, the pastures and hay ground had not recovered from the previous month's harvest. We scurried around to locate any available supplies of hay and were lucky enough to find a couple of sources. In the first week of July, we began feeding the flock purchased large round bales on one of their pastures. Now, in mid-September, they have yet to do any further grazing in a period in which they normally could barely keep up with the pasture growth. There will be no second cutting of hay. In late August and early September, we finally received rain in sufficient quantity to at least soften the ground and put a green tinge to it. If we are lucky and receive more rain, we may finally have some pasture to graze for a couple of weeks before the hard frosts of October.

Depressing Realities

It is easy to sense that the sheep are not happy with the situation. It is also easy to understand their dissatisfaction. Eating the same bales of medium quality hay in the same pasture day after day, when they are accustomed to fresh green pasture, cannot be very pleasant. For me, it has taken away much of the pleasure of caring for the flock. Gone are the daily morning walks to various corners of the farm to set up new fencing. With that is lost the best hours of the day when the morning's silence is only broken by bird calls or the distant mooing of the cows down the road. Also lost has been the joy of accompanying the flock to and from each day's location. All this has been replaced by the drudgery of moving half ton hay bales to the same bare, brown, dry pasture week after week with no hope of improvement. Even the mature hardwood trees that surround our pastures, the maples, birches and beeches had begun to die by mid-August. Most depressing is the thought that should this type of situation continue/repeat itself next year, we will not be able to continue farming as we know it. It is hard to imagine our lives without the sheep and their lambs.

By mid-August it, was clear that I needed a change of scene and attitude in order to survive the rest of the year. Last year I had planned a two week trip to visit old friends in Denmark. The trip was cancelled at the last minute due to our need to purchase a replacement tractor, coupled with the opportunity to add a significant group of Corriedales to our flock. We decided that I should resurrect my trip plans from the previous year. We had been able to purchase enough hay to feed the flock into mid-September. I could in advance set up enough of the large round bales that Gretchen would not have to deal with the mechanics of feeding the flock for two weeks. I would travel to Denmark in the hopes of refreshing my outlook and attitude.

A Return "Home"

What follows is a very brief account of some of my experiences in Denmark during the last week of August and first week of September. I will try to limit this accounting to that of the shepherd's perspective, but I need to at least give some background into why Denmark was my destination.

Forty two years ago, after graduation from high school, I had the rare opportunity to live for thirteen months in Odense, Denmark attending an advanced secondary school (gymnasium) as an exchange student. While there I lived with a Danish family. I was treated as one of their sons. At the gymnasium I was just another student, albeit one who began the year with virtually no knowledge of the language. It was an exhilarating and exhausting experience. In retrospect it was the most defining thirteen months of my life and it has left a permanent imprint. By the end of my stay I was approaching fluency in Danish and felt as much at home there as I did in my birth country. I became as close to my Danish family (Far, Mor and five brothers) as I was to my biological family. Over the years we have remained in touch, but I have managed to return only rarely. About thirty years ago, I returned with Gretchen so that she might meet my family and see something of the country. It was then only two years ago that I was able to again return. Serendipitously, this summer one of my brothers planned a family get together. It was the first time that all six of us had been together in over 41 years. It was a happy, emotional time. Over the next twelve days, I spent time with my brothers and their wives. In between I traveled along narrow country roads, visiting familiar haunts and seeing parts of the country where I had never been. During my original thirteen month's stay, I lived in a city and had very limited exposure to agriculture and no interactions with shepherds and sheep. This time I made a purposeful effort to visit a couple of locations with interesting flocks. In between, I would stop to look at flocks whenever I happened upon them during my rural travels.

My time was spent exclusively on the Jutland peninsula (Jylland) and on the isle of Funen (Fyn). Due to time constraints, I was not able to visit the eastern islands of Denmark; therefore my sheep perspective may have a geographic bias. Many of my observations about the sheep and their environment will also have been biased by the weather. Climate change has also been drastic in Denmark. This summer they have been clobbered with more rain than literally ever before in history. I saw lush green pastures, often partially flooded. The sheep (and I) were often wet, which tends to alter their wool's outward appearance.

Denmark is not widely known for its sheep, both in terms of numbers of sheep or native breeds. The vast majority of flocks are small, sometimes a supplement to larger dairy or hog farms or often stand alone mini-flocks. The largest flock numbers around 1000. Most flocks may be from 20 to 40 ewes. There is a heavy emphasis on meat breeds and production but there is a small fiber segment. While Denmark, like the rest of Scandinavia, has a strong heritage in the fiber arts, it is my impression that there is a rather small community of active handspinners. I am not sure whether that is the result of limited availability of quality spinning fleeces or whether the small number of handspinners (and therefore the limited demand for handspinning fleeces) limits the number of fiber flocks. I was able to come upon a good number of quality boutiques offering finished wool and mohair products and to a lesser extent yarn. Nowhere did I find anyone offering raw fleeces or roving for sale.

Texels

In southwest Jylland, especially along the coast of the North Sea, I saw the greatest concentration of sheep. The primary breed was overwhelmingly Texel. This area is not that far from Holland and the origins of the Texel breed. Also, it is an area similar in habitats. The greatest numbers of Texels and Texel crosses that I saw were grazing the long, extended dikes that hold back the storm surges from the North Sea. It is an ideal grazing scenario; great expanses of open grassland which cannot be cultivated to crops. Many of these Texel type flocks contained a small yet significant number of colored sheep. This colored minority was almost exclusively pure black, but otherwise very physically characteristic of the Texel breed. It was also interesting that so few other color patterns appeared outside of the pure black. I came across some flocks that were obviously purebred Texels (probably high quality breeding stock). They were perhaps the finest Texels I have ever seen. In other cases, the primary Texel flocks contained some cross breeding. From my observations these crosses tended to be to either Suffolk (English type) or Gotlands Pelsfår (we know them simply as Gotland).

Gotlands Pelsfår

I was particularly interested in seeing a number of different races of the Nordic short-tailed sheep, many of which are not found in the U.S. The most common of these breeds found in Denmark is the Gotlands Pelsfår (literally the Gotlands Pelt Sheep). It is a breed which originated on the Swedish Island of Gotland. It is not to be confused with a much rarer cousin, known as Gute sheep, a more primitive horned breed. The Gotlands Pelsfår is thought to be a cross of the old short-tailed landrace sheep with Karakul and Romanov sheep brought to Gotland from Russia by the Vikings. Beginning in the 1920's, a project of intensive selection amongst those sheep was started which eventually produced the modern Gotlands Pelsfår. They were first imported into Denmark in the 1960's and now number over 10,000 in Denmark. They are known for their good mothering ability, multiple births and easy lambing. It is a triple purpose sheep raised for meat, pelts and lustrous wool. The modern Gotlands Pelsfår is a polled breed. The face and legs are free of wool and usually black. The fleece varies from light to very dark gray. It is lustrous, soft, and curly. It lacks the double coating of the more primitive races. They are shorn twice a year.

I was able to spend a couple of days at Pension Holm Mølle, a bed and breakfast located in the rolling hills a few kilometers north of Silkeborg. The owners, Dorte and Nils Kærn, are relatively new to both the bed and breakfast business and the sheep business. They have a flock of about 40 ewes which they raise for meat. The flock is Texel crossed to Gotlands Pelsfår. They added the Gotlands genetics to the flock because they have a reputation for easy, reliable lambing and mothering. I was not able to experience it, but many Danes also believe that the Gotlands adds a different flavor to the meat (a flavor that is apparently much in favor). One could also easily see the influence of the Gotlands genetics on the Texel fleeces; the wavy, lighter crimped wool caused many of the tighter Texel fleeces to seem more open. They also added a color element to many of the fleeces.

I had occasion to stumble upon a couple of Gotland flocks elsewhere. They are a handsome sight up against the lush green pastures, surrounded by hardwood forests. Perhaps most intriguing was a Gotland flock on the grounds of Holstenshus manor house in the south of Fyn. Not only did they have their own private lake around which to graze, but they also had a private half-timbered shelter, very much in keeping with the 18th century farm house architecture of the area. It was also a flock with a range of black to grays shades in the different fleeces.

Spælsau and Luneburger

I had hoped to get good close views of two relatively rare (for Denmark) breeds: Spælsau and Luneburger. Both are represented in the large flock at Lystbækgaard, in northwest Jylland, near Ulfborg. The farm and flock are owned by Berit Kiilerich. She had been the shepherd for a state owned flock that was used to manage and control the undesirable trees and shrubs which otherwise overgrow the native heath areas which at one time predominated large portions of Jylland. For financial reasons, the government got out of the sheep business. Berit was able eventually to purchase the flock and Lystbæk farm, thus continuing the grazing operation in the heath around Lystbæk. She has established a craft center at the farm, which now operates much as a cooperative or

226

association. Its purpose is to preserve the old, nearly forgotten, handcrafts and traditions and to pass them along to the farm's visitors. Classes are offered in spinning, weaving, felting, willow weaving and other old crafts. I had hoped to talk a bit with Berit about her flock. Unfortunately, I could only be there while she was hosting a group of women from an Icelandic agriculture association. She was busy with the group and their schedule. I was at least able to tag along for part of an afternoon. Her flock is around 400 ewes, actually divided into a number of smaller groups, some mostly Spælsau, and some Luneburger Sheep.

The Spælsau sheep is from Norway and may also be know as the Old Norse Sau or Colored Spælsau. It is now rare in Norway where less than 1,000 ewes are registered. It is felt that it is the same type sheep from Viking times that was eventually established on the various North Atlantic islands. It is short-tailed, lacking wool on head and legs. It can be found with and without horns. The fleeces are double coated with a soft fine undercoat and a long, lustrous outer hair. They come in a variety of colors and combination of colors. Often the outer guard hair and soft undercoat are in different colors. The wool lends itself to spinning and felting. It is a breed that is well suited for habitat maintenance i.e. they aggressively control tree saplings and bushes, making it well suited as a grazer on the heaths of Jylland.

I was able to see a small number of the Spælsau ewes up close. The color and pattern variations were fascinating, especially for someone with a Corriedale perspective. (I could only dream of seeing some of the browns and chestnut shades in our fleeces!) We visited a flock of ewes and lambs in a steady rain; it was easy to see the protection of the double coat at work. Later, I came across one of the ram/ram lamb flocks which had been recently shorn. Without the longer guard wool overlaying the rest of the fleece, the colors were all the more intense.

Her flock of Luneburge Heath Sheep were at a distant location and therefore not accessible. The breed actually originates from the large Luneburge heath of northern Germany and are also known as Luneburger or Graue Gehörnte Heidsnucke or Gray Horned Heath Sheep. Apparently today they have become something of a national landmark for the German heath area, where wandering flocks with their shepherds are supposedly a common sight on the heath. The breed may originally stem from the wild Muflon sheep. It is felt that at one time the breed was found over much of what is now Denmark, thanks again to the Vikings. It, however, did not do well in captive environments and gradually disappeared. The current Danish population was imported from Germany in the mid-1930's. At present there are less than 1,000 ewes in Denmark. It is a relatively small sheep, with a gray pelt consisting of a fine wool undercoat and darker long guard hair. The wool is naturally shed in June. Both sexes are horned. I was at least able to get close to a small group of ewes and lambs at a reproduction Viking village near Ribe in southern Jylland. The uniform gray/black coloration was quite a contrast to the colorful Spælsau. I was also able to find a small supply of their wool. What a dramatic difference between the kempy outer fiber and the soft under wool!

Danish Landrace Sheep

By going to a couple of "living" agricultural village museums, I was also able to see a few Danish Landfår (Landrace) sheep (plus a very impressive Danish Landrace goat buck!). They are also known as Klitfår (Dune Sheep), based upon their origin in the West Jylland dune areas. It is a hardy race which could survive in a harsh environment where other breeds fail. Because they later spread over the entire country the "Landrace" rather than "Dune" reference prevailed. There were early efforts in the 18th century to develop it as a homogeneous breed. The introduction of foreign breeds (especially English) to Denmark tended to eliminate the breed from general use. There are now estimated to be only about 1,100 Landfår remaining.

It is a long-tailed sheep. They are polled and white in color with the head being either light, dark or speckled. Legs and face are free of wool. The wool is otherwise dense and appears to be of good quality. At one time, they were reported to produce three to four lambs, but with their drastic population decline and loss of genetic diversity, it is now normal that only one lamb is produced. It is an altogether different sheep than the old Nordic short-tailed breeds. Superficially, they reminded me in appearance of the usual first generation offspring of the cross between white face ewes with black faced rams. I was not able to get a feel of the raw fleece, but was able to purchase a couple of skeins of mill spun yarn. Off-white in color, it feels rather rough and coarse to me, but that may be biased by the millspun nature of the yarn. I will wait until we can wash and work with the yarn before I can reasonably comment on it. Sometime, I will try weaving something from it for a better sense of its character.

A Return to My Own Home

In the end, it was very good to visit a place I look upon now as my second home. I had time to visit my adoptive family and renew very deep friendships. I was able to visit some very special places that hold wonderful memories. It was pure joy to begin to feel comfortable speaking and understanding Danish as I once did. To be able to combine these very personal experiences with the excitement of exploring a lovely, agricultural environment and communicating with a very warm and friendly group of people that typifies the Danish society was very satisfying. It was also good to return home to Gretchen, the flock and farm. The problems brought upon us by climate change here, in Denmark, and elsewhere were not corrected. In fact they are all the more evident. But I can at least smile a bit more for the moment.

A Few Words from the Big Barn

By Nanoo Nanoo,
editor for Baa Baa Doo Press

My name is Nanoo Nanoo. If you have managed to struggle through this book to this point, you may have guessed I am **not** the guy who usually writes here! You may be asking, who the heck is writing this? Let me give it a shot and then perhaps you will better understand the situation. I am one of the sheep who live at Whitefish Bay Farm. I am not that old, bearded, shepherd guy who usually writes here. He and his wife, the nice lady, live up at the other barn. (He calls the place "The House"…kind of a strange name, but we try to humor the guy.) It being the dead of winter, I currently am spending my time in the big barn with my family and my buddies. When we heard that there was going to be a book coming out about the farm, we figured that it was time that we had some direct, **unedited** input. So far you have only heard from the old, bearded shepherd. We need to make absolutely sure that our side is being fairly and honestly represented.

In order to achieve our goals, we had to hack into the old guy's computer. It was not easy, but then we are a lot smarter than humans give us credit. How'd we do it? Every year when lambing time rolls around, the old guy sets up a baby monitor in the barn so

231

that they can listen in on us without spending all their time down here. For a few evenings, when the people were sleeping, one of my geeky cousins, Queso, and I re-wired the monitor so that it also acts as a Wi Fi antenna and terminal. With that installed and with a bit of educated sleuthing, we managed to hack into the old guy's computer. So here we are with access. Once we were successful, the bunch of us set out to keep the world up to date with the truth about how this place really operates. If the old klutz does not notice, we may get a few good chapters inserted into the publication!

In the distant past, the old guy had enlisted my help in writing a few newsletters. A few years ago, I even wrote the Christmas letter from the farm. After agreeing to wear a very hokey Santa hat, I even got my picture published with that holiday greeting. Since I now have some literary background, the rest of the girls here in the barn elected me to write the chapters. If I am lucky, I will endeavor to introduce you to at least some of the members of the flock.

I suppose that I should start by formally introducing myself. The name the shepherds gave me is Nanoo Nanoo. That means that I was born in the "N" year, which most of you know as 2002. Each year the shepherds use a different letter of the alphabet to name all the lambs. They claim it makes it easier for them to remember how old we all are. I think it is just because they are numerically challenged. My mom is Mindy (she had a cousin named Mork) and my dad was Mercury. The word is that I was pretty cute when I was born.

Nanoo Nanoo

I have tried to get a picture of me with the old guy into the book, but it seems that the old guy is too cheap to print it with photos. I have a picture of me with him when I was just a few weeks old. If you could see it, you would notice how he tries to hog the center of the picture! Back then my fleece was a lovely light, variegated gray color. Now that I am older, I have become a much lighter gray. A couple of years ago, the nice lady kept my fleece for spinning, rather than sell it. I think she is still spinning and knitting with it. The old guy may even be weaving with some of the yarn from my fleece. Over the last six years, I have given birth to seven lambs. The shepherds tell me I do a very good job in labor and delivery. I know that I am an excellent mother. I am especially proud that all of my daughters have remained with the flock. In just a few months, I will be lambing again. However, I am not telling whether I am going to have a single or twins! I tend to be very friendly when B&B guests come around to visit (what the heck, someone has to be nice around here!). I have also taught my daughters to behave that way. We figure we need to help the shepherds out any way we can with public relations. I will admit that they take pretty good care of us.

This should wrap up my first independent venture into publishing. It is a bit on the short side, but I am just getting the hang of it. I should be back in a while with some truly interesting stuff about some of the other members of the flock.

Disrupting the Routine

Anyone who has been around a flock of sheep for any length of time knows that they are creatures who thrive on routine. They are most content when each day evolves much as the previous days have done. For our flock, security is found by going out to pasture at the same time each day, using the same route, through the same gates, finding a new paddock with new grass which is just as fresh and tasty as yesterday's. Using a different route is more stressful for a few days. One can sense an uneasiness going through a gate that has not been passed through in a long while. Similarly, the flock comes to know when it is time to come home, even if there is still ample grass to be grazed. After all, they are quite content with the established rhythms and patterns of the day.

What we, as shepherds, often fail to recognize is that we too are just as comfortable with the daily routines that we have established for the flock (and for ourselves). Often we do not even recognize how tuned into the routine we are, until it is changed or disrupted. I even suspect that our discomfort with any change is often noticed by the sheep. In that situation, it can also become a disruption in the flock's routine. Working with the flock each day, in a sense, makes us part of the flock. Perhaps, we may just not be as observant as the sheep.

On our farm such disruptions in routine seem to be much more evident when one of the two of us is absent. I suspect this is reflected in every flock. In most situations, being a shepherd is being part of a team. The classic case is a single shepherd who works his or her flock with a herding dog. In a larger range flock, the same team will probably include more than one dog and at least one horse. In other scenarios, the team roster may include one or more guard animals, be they dog, donkey, llama or some other animal. Other variations may include additional people, be they family members or employees. In our situation the "team" currently consists of two humans: my partner and wife, Gretchen, and myself.

Remove one or more members from the team, even briefly and the entire dynamics of the flock's daily routine is likely to change, sometimes dramatically. On our farm, we become especially aware of these dynamics if either Gretchen or I am absent for a day or longer. Even though we live in a rural, agrarian area, it has become increasingly difficult to find qualified or interested outside help to temporarily replace one of us. The farming population here (as it is in the rest of the country) is aging. There are fewer young farmers replacing the old. As such, there are also fewer farm children around. In the past, they could be a good, reliable source of temporary help. In the few remaining young farm families, the kids have become a vital cog in the farm's operation, so much so that they rarely have the time and/or opportunities to help others. The same applies to their grandparents. They too are remaining active in their family's farm operation long after they "retire".

Faced with this scenario, when Gretchen or I leave the farm, we try to plan the period so that whoever remains here is, hopefully, not faced with insurmountable problems. Obviously, it is our hope that the change in our routine as shepherds will also have as minimal a disruptive effect on the flock as possible. Inevitably these times are also when Murphy's Law or a variation of it often comes to the fore: If something is going to go wrong, it will wait until one of us is gone. If we were ever able to break their code, we would probably discover a similar law in the flock's Ovine Book of Conduct: Whenever one shepherd is gone, put the remaining one to the test.

We first became aware of this condition the very first time I had to leave the farm, the flock and Gretchen for a goodly length of time. Many years ago I had to travel to California due to a family health emergency. This trip occurred in the dead of Wisconsin winter, at a time particularly suited for anything to go wrong. At this time of year the flock is being feed hay in the barn; all the ewes were about a month away from lambing. The actual daily care of the flock should be very routine. I had been gone for over a week when our yearling ram, Haakon, decided to challenge the status quo. He was still not large enough to reside in the same pen as the adult rams, so he had his own pen next to them in the back of the barn. Gretchen has never had a great empathy working with any of the rams, including, especially, Haakon. I suspect that the rams sense it and at times try to take advantage of her. (The "guys" and I get along; they seem to know that as long as they do not mess with me that they will not be in too much trouble.) Haakon decided, after a week into Gretchen's watch, that he was tired of life in his own pen. One morning Gretchen found him in with all the ewes, happy as could be. How he managed to unlock the gate or jump over the fence to get into the main pen is still a mystery. She was, not surprisingly, unable to cajole, tempt or force him into returning into his own pen. She also did not feel that alone she could catch and muscle him back home. Of all days to pick, Haakon had chosen the most difficult time to find <u>anyone</u> to come to Gretchen's assistance.

It was Super Bowl Sunday and the Green Bay Packers were playing. People unfamiliar with northeast Wisconsin cannot imagine how much this part of the world came to a standstill that weekend. She could not find any help and eventually Gretchen gave up. She hoped that if there were any open ewes left that they would not cycle while Haakon was with them. His capture and return to bachelorhood would have to await my return in a couple of days. Of course, none of this history had been related to me until I got off the airplane and we were headed home. It was late at night by the time we arrived at the farm, so it was decided not to disrupt the flock. We waited until morning to deal with Haakon. When we came down to the barn the next day, it was as if life had instantly returned to "normal". I walked over to Haakon's pen; he came over to visit. I opened the gate and much to Gretchen's disgust Haakon walked in all on his own. I do not know how or why he did it, but it was apparent that the routine had been re-established.

It took another decade for history to repeat itself. I was traveling in Denmark this fall. The ewes were still out on pasture; due to a shortage of grass, the rams were confined to the barn. One morning Gretchen found that someone had opened the gate between the ram lambs' and adult rams' pens. This gate is a "new and improved" version, much more difficult for anyone to open than the one with which Haakon had to deal. (It is a gate that not anyone in our large animal veterinary practice seems to ever be able to open!) The result was that three ram lambs and seven adult rams were all mixed together moving from one pen to the other. To Gretchen's credit, she eventually managed to get all of the ram lambs back into their pen and all but one adult, Oz, into their enclosure. Oz stubbornly refused to return. The decision was made: As long as Oz did not bother the little guys, he would just stay put until my return. As it turned out, they got along pretty well. It seemed to lower the

238

degree of conflict among the rest of the adult rams. So for the time being, Oz stayed put, even after my return.

Eventually, later this fall the rams returned me the favor. Gretchen was off to Michigan for a little over a week to attend the Spin Off Autumn Retreat. One morning, upon coming to the barn to take the ewes out onto pasture, I found that the three ram lambs and Oz had opened the "unopenable gate" and moved into the main barn where the ewes resided. There they had found a couple of extra hay bales of which they had managed to make a complete mess. I thought, at least, that they had not managed to get into the pen with the ewes. I was able to get them back into their own quarters, but it was then that I discovered that one of them, Ukiah, was missing. I had just assumed that he was still in the ram lamb pen. Back in the main barn I located him, seemingly quite pleased with himself, literally surrounded by over a hundred females. He had obviously jumped a fence. How long he had been there was only conjecture. Understandably, Ukiah was not particularly happy when I caught him and moved him back into the ram quarters. That maneuvering got the rest of the rams so excited that they bashed open the gate between their two pens. I was then forced to sort the bunch out without an extra hand to open and close gates. I eventually managed (with the help of a bucket of grain in a feeder in one pen) to almost get every one into the adult pen. From there I worked at getting the ram lambs back into their pen. In the process, Oz found himself back with his old adult buddies. So, after two separate solo shepherd performances, all the rams were back where they should be. As a post script, we watched all of the ewes that we put into breeding groups a week after Gretchen's return. All but one was marked by the ram in their group. At least, as we approach lambing time we "only" need to watch for an unexpected pregnancy for the one unmarked ewe, all of our retired "old ladies", and the ewe lambs we opted not to breed. Until we reach shearing, a couple of weeks prior to lambing, we will probably not know if Ukiah's night with the ewes had any lasting consequences. If we have any unscheduled lambings, it will be easily attributable to my time alone with the flock.

As we think of these situations, it is fascinating how many similar "unscheduled" events have occurred while one of us was absent. For example, included on the list are the following. A waterline far out into the pasture waits all summer to break until I am away. A spool of portable electric fencing becomes hopelessly tangled into knots one morning while Gretchen is moving fence alone. A ewe develops a serious eye infection while I am here alone. Not only was she one of our least trusting ewes, but after the first day of eye drops, she was almost impossible for me to simultaneously restrain <u>and</u> apply eye drops for another week.

The moral of the story is that if one dares break up the shepherding team, even momentarily, something is bound to get goofy or messed up. Appreciate the other members of your shepherding team whoever the members are. Give the herding dog extra pats and lots of praise for a job well done. If you have a co-shepherd, give them some time off, and when they return thank them even more for their contribution and sharing of the work. Give them lots of extra hugs. It will be very good to get back into the routine.

Finally Raining!

What a difference a year can make! Twelve months ago today, June 8th, I began cutting hay for the year. From that day onward I was able to cut and bale our hay fields continuously until finished. Normally making first crop hay is an "iffy" game in Wisconsin. One is often racing to cut the hay, get it properly dried and baled before the next summer rains move through. Last June was different. Even before we finished with hay and had it all stacked in the barn, we realized that we were in the midst of a drought, the likes of which we had not experienced since we began farming in 1983.

Today I sit and watch the rain come down steadily. It is a day for indoor projects. It is now so wet that the sheep have been allowed to stay in the barn to munch on remains of baled hay from last year. The weather has been like this, off and on, for much of the spring and into early summer. In years prior to "The Drought", these conditions would have started to bother me. When will I be able to start cutting? The grasses in the field are starting to head out; quality will decline. How will we ever get things done?! After last year, we will be happy to just cut sufficient quantities of hay to make it through the winter. In the meantime, it is gratifying to see the pastures grow faster than the ewes and lambs can keep up with them. We hope it is a long while before we have to again feed purchased large bales on what was once green pasture. It is an experience that was mind numbing for me as a shepherd and grazier. It was also an experience which the sheep obviously found unpleasant compared to grazing. The sight of them returning to lush green pasture was extremely gratifying from our perspective. Every twitch of their body language seemed to spell sheep happiness.

The long term influences of last year's drought are still to be seen everywhere. It was dry enough that the buds which set on the cherry trees in the fall were compromised. That, coupled with a return to colder and more seasonal winter temperatures, resulted in a nearly total loss of viable cherry blossoms this spring. The local commercial cherry harvest this year is already assured of being a total loss, despite the current ample moisture.

It will prove interesting to see how well the legumes (alfalfa, clovers and trefoil) in our hay field have survived and regenerated. At present, my guess is that we will see some losses, despite the ample moisture. We have purchased a good supply of red and ladino clover seed in anticipation of such losses. The plan is to broadcast the seed, using a hand-cranked spreader, in each pasture a day or so before the sheep will enter the area. If all goes well and the ground is still moist, the sheep will tread much of the seed into the damp pasture soil as they graze. Perhaps by fall, and definitely by next spring, we should see nice legume regeneration. The added bonus will be the additional nitrogen fixation preformed by the legumes (a distinct plus considering the soaring costs of commercial fertilizers!). If the pattern of repeated rain continues, it should allow us to reseed most of the pastures just prior to the sheep entering them for grazing.

Such small scale rejuvenation efforts do not translate well to larger tracts of pasture, like our 40 acres of hay ground. First, it represents a huge area onto which to hand broadcast seed.

Secondly, there will be no follow up by the sheep to tread-in the seed. We had planned originally to attempt no-till drilling the clover into the pasture as soon as the frost left the ground. Slow planning on my part resulted in the delay in the arrival of a small no-till grass/legume drill. Compounding the delay and late delivery of the drill was an unanticipated amount of assembly work smack in the middle of lambing. Needless to say, the lambs got top priority and the drill was temporarily neglected. Once it was ready, the hay field had already started to grow too much to allow for any success. Hopefully, we will now be ready for this experiment, either in fall, or for sure, next spring.

Last winter's weather also had unexpected consequences for us. It was good to see a more normal snow pack for much of the winter. (The southern portion of Wisconsin had record setting snowfalls. I am sure that many folks in that area wondered if they ever would be able to see above the drifts.) A number of snows were accompanied with ice storms, either before or after the snow. The usual consequence is a layer of ice, just about everywhere. Our old dairy barn was built into a hillside. For the sheep to get outside, they must descend a moderate incline. When the incline is covered with two inches of ice, the sheep are essentially stuck in the barn until the next thaw. Conversely, it is impossible in these conditions to get our small skid-loader up the same slope and into the barn. Normally, we scrape out the barn about every three months. In between, we let the pack of bedding build up. The presence of a concrete floor in what was the milking area for the cows is a disadvantage for us. We must bed heavily enough to avoid a moisture build up. We normally clean the barn completely in fall prior to bringing the main flock in from breeding and then again three months later prior to shearing in late February. In that way, we can have a soft, dry and comfortable bedding build-up in place in time for lambing in March. This year the ice proved to be our undoing. Not only could the sheep not get out, but by shearing time the skid-loader could not make it up the incline to get into the barn. The two inches of ice on an uneven slope seemed impenetrable. We were faced with the reality that, in February, the barn would not get scraped out and that lambing would therefore proceed on a deeper than normal bedding pack. It would also mean that any clean out would have to wait until the flock, complete with lambs, started grazing in May. So now, as time and weather allow, "Eunice" the trusty old skid-loader, and I are plugging away at barn cleaning. It will certainly result in logistical log jams down the road. The normal amount of bedding and manure generated from November through February has not been sitting in the manure storage where it usually decomposes nicely enough to turn over and make room for the spring clean out. Someplace this fall I will be looking at a prodigious amount of partially decomposed manure which will need to be spread on pastures. At least it makes for very good fertilizer!

Throughout all of the unusual variations on the otherwise normal themes which play out on the farm, lambing, at least was one of the most consistent. We intentionally did not breed quite as many ewes last fall. This decision was, in part, based upon our growing population of healthy but older ewes. We no longer breed a ewe after age eight. Our geriatric set still produce quality fleeces and tend to be relatively easy to care for. It is good to have a bunch of old, long term friends in the flock. At the opposite end of the age line we are no longer breeding our lambs at age seven months. Obviously, this change results in one (and possibly two) fewer lambs each ewe produces over her lifetime. It is, therefore, a reduction in her overall financial productivity. It has, however, tended to produce fewer lambing problems which we associate with first time moms. These older first time mothers are more physically mature. They are, in most likelihood, emotionally more mature. (How one measures that I am not sure!) Thus, they are better mothers producing lambs which tend to be slightly larger. Their lambs also grow better than when they give birth at twelve months of age.

This year's ewes produced 109 lambs, all of whom were healthy and robust. We lost one lamb due to an apparent injury, probably caused by crowding at a feeder. So far the average rate of gain for the lambs is well ahead of last year's average. Ewe health has been remarkable. I do not want to think about the complete lack of any noticeable signs of mastitis for fear that one will appear. One is tempted to wonder if there is any correlation between the lack of health problems and a change in the bedding conditions in the barn after lambing.

Fewer health problems for ewes and lambs and fewer climate related stress problems for sheep and pastures, when combined, produce another large benefit: less stress on the shepherds. When compared to last year, this spring and early summer has been a much more rewarding day to day existence for the two of us. As I finish writing, the rain has let up enough so that I can see to the far side of the five acre pasture outside the window. A doe White-tailed Deer has just appeared with the first fawn we have seen this year. Over the last few days, a pair of Sandhill Cranes has been either stalking through the 40 acre hay field beyond or we have seen them on the edge of our cedar/ash wetlands (which are again wet!). The chances that they are nesting in the wetlands are remote, but the prospect is exciting. It will be nice to be worrying about spotting a fawn or crane chick ahead of the haybine rather than worrying about whether there will be enough hay.

A Postscript

In the few days that have passed since I started and finished my "on again off again" efforts to write this piece, our rains have continued, also in an "on and off" fashion. For those who will read this a couple of months from now, I should offer an apology. For us, the rains have been timely but not excessive. For others not so lucky, in southern Wisconsin, much of Iowa and parts of Indiana, the rains have been excessive and the resulting dramatic, flooding extreme. I offer my greatest sympathy and hope that you do not misinterpret my joy over the ongoing rain. Similarly, to those in the western U.S., New Zealand and Australia who continue to experience drought beyond belief, our hearts also go out to you. May all of you find the climate middle ground we so desperately need.

Sheep Barn

Mid September 2008 has arrived and the season is rapidly changing. The summer has been very pleasant and productive. All but four of this year's lambs have left for new homes or for market. Our ewe flock is at its projected size for breeding. The ewes and rams will be in their various breeding groups in exactly one month. Ample rains have allowed us to harvest enough hay to fill the barn for winter and beyond. Pastures have stayed green and productive despite a brief dry spell in mid and late August. The flock has continued to graze from spring through the summer, interrupted only for a couple of days by thunder storms and heavy rain. Currently, the ewes are finishing off a section of the hay field. The barn was already filled with hay so the girls have gotten to harvest the last five or so acres of what would otherwise have been a second cutting of alfalfa and grass. Hopefully, it should be a good nutritional boost before breeding.

Migratory birds are on the move. The Canada Geese are forming larger flocks, a few of which are flying much higher and are pointed in a southerly direction. Sora Rails are flushing out of the deeper portions of the pasture each morning as I move the temporary fences. These little "explosions", in the otherwise quiet pasture continue for about a week and then the Rails are gone. "Our" colony of Barn Swallows pulled out one day in early

September. Just prior to departing, they would spend the early morning sunning themselves on the top wire of the fence to the most westerly permanent pasture. Somehow they knew that the sheep would arrive in that pasture soon after sunrise, stirring up the bugs just for the swallows' breakfasts. Near their departure I tried to count the number of swallows, but it was an impossible task. My best guess was that they numbered at least 125 to 150, all of whom certainly were born in our barn. It is probably the largest number our farm has yet produced. Now they have departed; I already miss their happy "conversations", joyous flights and their companionship on the pastures and in the barn. Their return next spring cannot come too soon!

The rams are preparing for fall and breeding. The adult rams have become especially smelly. Their noses are again becoming wrinkly. Their tolerance for each other is now also strained. The ram lambs are mimicking their fathers and uncles, sometimes in an almost comical fashion. However, from their perspectives, they are quite serious. The last couple of weeks have been difficult for all the boys. Much of the time they have been cooped up in the barn and forced to eat baled hay rather than graze fresh green grass. The reason for their confinement has been an on and off project to replace the barn roof. Much of the time, the lift equipment needed to get up to the top of the barn has had to operate in front of where the rams would normally exit and enter the barn on their way to and from their pastures. The presence of the roofing crew and their equipment and materials has effectively blockaded the rams in the barn on weekdays. When they are able to get out to their pastures, their joy has been obvious. The seven ram lambs have always run down the hill, kicking up their back legs, twisting in mid-air and continuing on at full speed. Normally, the adult rams are much more reserved each morning; a sedate trot usually will suffice. But after five or six straight days in the barn, the old guys cavort down the hill just like the lambs. About the only difference is that that cavorting usually includes a few obligatory head butts before they actually make it to their pasture. As of tomorrow, the rams will be able to get back to their normal routines. The roof will be finished this afternoon!

New Roof

Unfortunately, every once in a while, a major maintenance need makes itself evident on a farm, especially if it is populated by old buildings. This last winter and spring it became apparent that the barn needed to be re-roofed. After each major storm and especially those with strong winds from the south and/or east, we would find large chunks of asphalt shingles on the ground around the barn. Twenty four years ago we had to replace the roof shingles after a major hail storm. In the subsequent years, additional hail and wind storms took their toll on these once new shingles. As it was, the shingles just about made it through their expected 25 year lifetime. By now, there were enough small cracks and holes that the heavier rains were leaking into the hay mow. With the shingles starting to blow away, it was definitely time to put on a new roof, especially with the mow now full of hay.

After a lot of research, we opted to install metal sheeting with standing seams, rather than replacing with the same asphalt barn shingles. The initial investment would be greater, but the longer term benefits should offset the initial cost. The difficulty we had trying to find a contractor to apply asphalt shingles is a sign of the changing times in local agriculture. Twenty four years ago it was not a problem. I would guess that the declining number of large, old barns, coupled with the logistical difficulties of reaching all parts of the roof, have contributed to the decline in the number of roofers who will do such a shingle job. In the spring, we were able to locate a contactor who specializes in nothing but standing seam metal roofs. He had so much work state wide that he could not promise to begin until late July. In actual fact, the work began a month later than that. Luckily the existing covering on the roof held up until then!

One of the significant down sides when getting a new shingle roof is that the old roof usually has to be removed first. No matter how careful the roofers are, the resulting number of stray old roofing nails on the ground all around the barn is a big deterrent, especially when much of the areas around the barn will either have sheep walking on them or tractor and wagon tires rolling over them. The cost of removal and disposal of the old roof would be significant. However, as long as there were two or fewer layers of shingles, the metal roof could be applied over the existing material. Consequently, no great amount of dangerous material ends up on the ground. The roof also comes with a 50 year guarantee. Presumably, this one will be the last barn re-roofing that I will have to oversee. Visually it looks much different. It will take a while to become accustomed to it. So far, however, we like the new look.

Evolution of Our Barn

With the new roof, our barn has had three distinct types of roofing in its lifetime. Built in 1906, the barn began with a roof of cedar shingles, undoubtedly locally cut and milled. Those shingles are still there, albeit in not great shape. Subsequently, there have been three definite applications of asphalt shingles, probably spanning the period from the mid 1920's to this year. Now it sports a spiffy, charcoal gray metal covering.

250

The barn, as a whole, has undergone major transformations in it lifetime of 102 years. Much of its original structure and makeup is still evident, but some of the changes have been dramatic. The farm was originally a diverse dairy farm, which also housed pigs, chickens and working horses. Interestingly, the farm was never a home to any sheep, until we came along 77 years later. For many years (probably into the 1950's), horses were the only source of power to work the farm. There are signs of where they were stabled in the barn, with the cows. They occupied the east end of the lower level, an area that is now where we shear the flock. The wooden window frames at that end were well chewed as only a horse seems capable. The large wooden poles on which the horse harnesses hung are still imbedded in the walls at this end. When the horses left the farm, their area became a calf raising pen.

The main area of this lowest level of the barn was devoted to the milk cows. It is easy to detect three distinct expansions to the milking area, which reflects the gradual increase in the number of cows milked. In the far northeast corner was the bull pen. It showed the signs of wear and tear one might expect from its occupant. When we started our flock, the bull pen became the ram pen. Until we removed the pen with our "improvements", the rams looked like pretty tough characters locked in those quarters.

Originally the floor was dirt. It was transformed by poured concrete at least twice for the cows. The first pour included concrete feed mangers. The second, later pour included a gutter with mechanical barn cleaner. Prior to that, the manure was shoveled into a cart suspended from a track below the ceiling that ran out the door to where it could be dumped outside. When we started, the track was still present but the cart had disappeared. Our big transformation was to 1) knock out the concrete mangers and the stanchions in front of them and 2) remove the barn cleaner and fill in the gutter with concrete. The mangers served no use with the sheep and took up an amazing amount of floor space. The space they occupied on the western half of the barn is now where we set up our lambing jugs each spring. That space provides enough room for 12 jugs when we are at our peak period of lambing. The smooth concrete floor provides a good surface for cleaning the barn with a small skidloader (those who know her call her "Eunice", the Uniloader). Obviously, the concrete is not very permeable, so moisture can be a problem for us unless we bed heavily.

It is the lucky sheep farmer who buys a farm that is complete with a barn that was either 1) purposefully and properly designed for a flock of sheep or 2) is able to build something from scratch that will serve both current and future needs. The rest of us take the best that we can find and hope that we can modify the existing building(s) for our own needs. We feel that we have been lucky in that regards. Aside from major interior changes on the milking barn level, our only major alteration was the destruction of the silo and its replacement with additional housing at the end of the barn. The silo was not original to the barn in 1906 (predating the common use of silage). It was a later addition. When we had it torn down, it became apparent that the door from the barn to the silo had originally been a window, which was then expanded. We had no use for a silo (we did not plan to make and feed silage). In addition, the concrete work for the silo was starting to badly deteriorate. The threat of very large chunks of concrete falling on us or the sheep was great, as was the possible ultimate unplanned collapse of the structure. In the process of having it torn down, the removal of the resulting rubble resulted in a large hole in the slope south of the barn. Slightly expanding the hole provided room for an addition to the barn. The rams now have what is usually their own private quarters. An unexpected added benefit was the improvement of airflow into the rest of the first level of the barn.

Our 1906 barn is a classic Wisconsin German bank barn. It was constructed with locally cut and milled timbers in a post and beam manner. The major hay mow floor beams (which run above the lower level) are nearly 48 feet in length and usually 10 inches by 10 inches. Today, if one needed to make a major repair, it would be difficult to find comparable trees that large in our county. Should we ever have to make such structural repairs, it would be difficult to find the materials needed and the craftsmen capable of replicating the construction. The barn is built into a hillside (hence a "bank barn") which permits accessing the second level, the hay mow, at ground level on one side. This layout facilitates the emptying of hay wagons without the need for elevators long enough to reach more than two stories when extended. Hay balers did not exist when the barn was built. It was designed so that a large fork could be pulled on a track suspended at the peak of the interior. The fork was lowered to the floor and pulled by a horse to the doors on the uphill side. The fork would then take a large bite out of a well stacked wagon load of hay. The process was reversed to move the hay into the recesses of the barn. The track and pulleys are all still in the barn, but the fork was lost before we purchased the farm (not that we wish to go back to that system of hay harvest and storage!).

Currently, it is relatively easy, but time consuming, for two people to fill each of the bays in the hay mow with small bales using a hay elevator running from the outside next to a bale wagon. Making and moving small bales is relatively labor intensive. Loading, unloading and stacking bales is at least a two person job if it is to be done efficiently. In this regard, our barn cannot progress into a more modern phase of hay baling which can save on time and labor: large round and square bales. We have on occasion had large bales custom made from some of our hay crop. Most of it was intended for sale as a surplus crop. What large bales we have used need to be fed outside. There are no openings large enough on the lower sheep level through which a large bale could pass. Even if there was a large enough opening, the ceiling is too low to allow such bales to be maneuvered. If we could somehow overcome all those obstacles, it would be difficult to store the large bales in our current hay mow and then transport them to the lower level of the barn for feeding. The combined weight of the large bales and the tractor needed to move them would be too great to trust the current structure of the hay mow floor. If they could be stored there, it is still a quarter of a mile to transport them around to the other side; any shorter route would be too steep. Short of building a new facility, we must be content to continue our current scheme of feeding small hay bales in the winter months.

For the present, it seems that we are doing just fine with our facilities. We can dream of what it could be like in an ideal scenario, but we can also be happy the way it is. Short of a catastrophic event, we are certain that the sheep and the hay have a good roof to protect them for many years to come.

Tales of Auger Repair and
Christmas Trees

After nine years of faithful service, the grain bin finally decided to give us problems. Since the unit was used when purchased, we have no idea how many years total service it has provided.

When we first started farming, our flock was small and, as such, required a relatively small amount of grain to supplement the diet for all our sheep. It was a simple task to drive to our local co-operative feed mill and pick up our order of a number of bags of our mix of corn and oats. Once back at the farm, the heavy bags were carried one at a time into the barn and down the stairs to be stored in large galvanized garbage cans. As the flock grew, the trips to the feed mill became more frequent and the number of bags increased. The logical solution was to install a bin on farm. It could be filled by a large truck from the mill and would be unloaded by running the auger which reached from the bottom of the bin into the barn. It was surprising how long it actually took us to see the light!

There are only three times in their year when we feel that the grain is a critical element the flock's diet. In late gestation, the ewes require the extra nutrition to maintain their condition, while also providing for the rapid growth of their lambs-to-be. Once the lambs are born, the grain fed to the ewes helps them re-build their body condition and assures that they produce an adequate amount of milk to support the nursing lambs. When the lambs are starting to eat a significant amount of solid food on their own, the grain assists in their early growth. The need for grain in abundance begins in late winter and runs close to the beginning of summer.

This winter, regardless of its age, the bin suddenly decided to refuse to deliver grain into the barn. As is usually the case with breakdowns, the weather at the time was difficult: single digit temperatures, snow with strong gusty winds. Since we were not able to find the problem, we got an installer crew to visit a couple of days later. By then, it had become yet colder. Balky augers in cold weather can be the dickens to deal with, but it could have been worse. The repairmen could not get to us sooner because they were dealing with a large, uncooperative auger used to empty a manure storage at a nearby dairy farm (not my idea of fun on a frigid day!).

The initial thoughts of the repair crew were that we would have to unload the entire bin before we could get at the problem. The prospect of four tons of grain on the ground with more snow being forecast was not pleasant. One final hope was that the auger could be pulled from the bin without opening the bin. If the problem was inside the auger, repairs could be made and then the auger pressed back into the bin. As we began the disassembly, it suddenly became obvious that there was a plug at the end of the auger that <u>no one</u> had expected. It was a condition the repair guys had never experienced. It was quickly cleaned out; the auger assembly reinstalled; the entire system was tested; everything was good as new.

The sheep have now had their grain for two nights. Peace and contentment again returns to the barn. The revolution has been avoided! With any luck, we hoped to get our Christmas tree cut before it was too late.

In Search of a Christmas Tree

Even after the auger was repaired, snow storms seem to roll through the neighborhood every other day or so. Saturday, the 20th of December, provided us with a bit of a respite. For the first time this winter, we had time to dig out our snow shoes and go for a bit of a tour around the edge of the main 40 acre pasture in search of a Christmas tree.

We have a large number of fast growing balsam firs on the edges of the woods. It is quite amazing how the little trees seem to spring from nowhere and then in just a couple of years they reach respectable heights. Unfortunately, they seem to love the relative lack of competition along the pasture edges. Before too long their lower branches are growing into the electric fences which eventually will result in short circuits in the fence and a resulting lack of protection for the sheep. Around the large hay field, as the firs continue to grow, they begin to shade out the hay. Every once in a while, the trees need to be removed or at least thinned. At this time of the year, we can at least think of the task not so much as "weed removal" but as selecting the "perfect tree".

We found a nice choice, mid-way along the south edge of the hay field. The tree that I had been keeping my eye on this summer as a good Christmas candidate proved to be much too tall. (Somehow they seem so much smaller when driving by on a tractor and concentrating on cutting hay!) Close by, we were able to find another candidate of more realistic size. None of these trees have been trimmed to grow lush and thick. At this location, they are also on the north side of the woods so they receive less sunlight and are a bit scrawnier. This was not to be a sophisticated tree, but it made up for it in personality.

257

The walk home with the tree, through deep snow, was a bit slower than the trip out. Even for a small tree, it wanted to catch the wind and act as its own rudder, not necessarily in the direction we wished to travel. At least, we could retrace our tracks home instead of having to break a new trail in the fresh snow. The track we leave behind with the snow shoes is always well defined!

Once home, we realized that whatever the tree lacked in elegance, it would make up for it with its own uniqueness. Because it is from our own woods, it is also a part of us. With its trip home with us the tree brought with it special memories of a hike through deep snow on a crisp winter's day. A day later and another snow storm has completely covered our tracks across the field. The tree and its memories will help make Christmas a special time for us.

Hope

By Nanoo Nanoo,
editor for Baa Baa Doo Press

It is snowing and blowing again outside. Since the wind is coming from the east, we have a few mini drifts inside the barn doors even though they are shut due to the storm. Nonetheless, we are staying warm and dry. I have returned; it is me again, Nanoo Nanoo. I have managed to squeeze another chapter into the book.

The grumpy old bearded guy and the nice lady are up in their private barn. I think they may be playing around with avocados, trying to dye our already perfectly beautiful wool some strange color. There is also a rumor going round the barn that they may be getting ready for the shearer. We look forward to getting rid of these heavy coats, but shearing days are a lot of work for all of us, especially with our lambs due to arrive soon. We have been discussing shearing over the last few years. As always, the old timers have so much to tell about what we have not yet experienced. These discussions remind me that I had promised to introduce everyone to the other members of our flock. It is only fitting that I begin with Hope.

Hope is now the oldest member of our group, in less than two weeks she will celebrate her 13th birthday. For someone that old, she is spry and sharp as a tack. All of us greatly admire Hope;

no one else in the flock has experienced life as she has. We never fail to enjoy the tales that she has to tell. Somehow, the shepherds (the old bearded guy and the nice lady) seem to admire her too. She is special!

Hope

Her mother was Candice, one of the very first lambs ever born on this farm. I am told that Candice was a very good mom. Unfortunately, she became sick when she was pregnant with Hope. Hope was born a triplet, but sadly her two brothers were dead at birth. Her mom never got better after Hope was born and she died a couple of days after Hope's arrival. As a result, Hope was raised by the shepherds. (That was back in the days when they really did not know much at all!) Because she was the first orphan in the flock, Hope got to spend the first couple of weeks of her life in the barn that the shepherds call "The House". To this day, she is the only sheep in the flock who has been up there for any significant space of time. (There are a few of us who have been there for a couple of minutes, but only because we were not feeling well. As a result, the others who were there briefly do not remember much about "The House".) Hope knows what goes on up there. Her knowledge makes her special. Sometimes, we even think that she believes that she might be one of those "people". When the weather is good and we go out to pasture, she is almost always the last out the door. I think she feels sorry for the old bearded guy and wants to make sure that he does not get lost on the way to breakfast.

Hope is not a very large sheep compared to most of us. I think she may have even shrunk a little in recent years. She has white wool (like my mom, Mindy) but her dad was the famous Cimarron (apparently a bit of a rogue). He was the first colored ram in our flock. When Hope has lambs, she can have either white or colored. She seems to love all of them just as well. It has been a few years since the shepherds asked Hope to have any lambs. She is now retired, which means she gets to graze, eat and snooze as much as she likes. She was supposed to be retired earlier, but she decided to have one last fling in the spring of 2005. Unbeknownst to the shepherds, she and a few of her retired girlfriends invited the rams over for a party one evening in April. Much to everyone's surprise she had a couple of lambs, Shucks and Serendipity, that September. Now she has decided that was enough. Hope has outlived all of her children, but there are many of her grandchildren in the flock, one of whom is that friendly ram, Nicely Nicely.

Because she has experienced so much in this world, she naturally has assumed many important functions within the flock. She makes sure the shepherds know how to get out to pasture every day. Everyday during the winter for many, many years now, she has gotten to taste-test each and every bale of hay before the shepherds feed them to us. What would we do without our own Julia Child? I suspect that she gets treats every evening because of her important functions. (I think she gets even more than I do! How she does that I do not know, because she is very calm about it and not pushy at all.) Even the grumpy old bearded guy gives her extra grain! I think you can understand why we feel that Hope is indeed special and why we all love and admire her!

Summer Shades in the Dead of Winter

Winter seems to have us firmly in its grasp. Fall is ancient history and springtime, green grass and frolicking lambs are still so far away. And, by the way, it does not help that it is also cold! Nonetheless, the first week of January can also be a wonderful time. Even more predictable than the return of Bluebirds in March, the seed and garden catalogs all seem to arrive. They are like a warm breeze out of nowhere. It is time to plan and dream of next spring's and summer's vegetables and flowers. Our favorite catalog is from Johnny's Selected Seeds in Winslow, Maine. It just arrived in the mail. Like most seed catalogs available today, Johnny's is available online. Yet somehow, having a printed catalog in front of you is akin to reading a good book, one you can pick up, read for a time, do a little summer dreaming and come back to later. The cover of Johnny's catalog this year is especially "warming": strings of red and orange hot peppers.

The arrival of the seed catalogs also has complemented the project that is nearing completion on my large Glimåkra loom in our studio. It is the fourth in a series of blankets woven with the same pattern but with different colors. They are woven in blocks of warp-faced and weft-faced twill. The original blanket in the series utilized various shades of our natural colored wool in the warp and white in the weft. The subsequent blankets have substituted for some of the gray shades with various naturally dyed yarn. The dyes for the last two blankets have come from plants from both our vegetable and flower gardens, or from plants which grow wild somewhere on the farm. The current blanket includes yarn dyed from artichokes (three different shades), purple bearded iris blossoms, carrot peels and the berries from buckthorn (two shades).

Weaving the current blanket is a gentle reminder of last year's gardens. In the case of each of the vegetables that we used for dying, the materials were merely left-overs. We had some of our greatest success growing artichokes in 2008; the dye stuff was merely the juices that remained after cooking the artichokes. The iris blossoms were collected after a strong wind storm knocked over most of the plants. We thought for sure that the dye color would be in the purple range, but surprisingly came out a bright yellow-green. The buckthorn is an alien, invasive shrub/tree, the spread of which we are trying to control. We can collect the berries without fear of endangering a native plant. In fact, collecting the berries will, in a small way, slow down the spread of the buckthorn. The purple berries yielded lovely shades of brown, a pleasant reward for trying to control an undesirable plant. One of the wonderful characteristics of natural dyes is that their colors always compliment each other (unlike many chemical dyes which can clash). As a result, the blanket is a warm mix of colors which, in its totality, is a pleasant reminder of summer.

The "summer shades" blanket is almost complete. There are only about two inches left to weave and, after that, there is the finishing work (tying the fringes, trimming, pressing and the like). It should be a good day for that type of activity. While it is bright, sunny and warm at the loom, it is cold outside. A couple of nights ago the temperature got down to -12.5° F (24.7°C) and today is not getting much warmer. We are coping well, as are the sheep. It is relatively warm in the barn (20°F) with no major drafts. The sheep are fully fleeced and are quite comfortable as long as they have plenty of clean bedding, they stay dry and have a bit more than usual to eat.

With the emphasis for us on indoor activity, it is time to think about the next project for the loom. Gretchen has dyed a good collection of wool in blue, red and violet shades using natural dyes from indigo, cochineal and logwood. Those dyed yarns plus some of our naturally colored gray yarns will most likely be the source for the next blanket.

Six Months of Avocados

Just down the road is another dye project. Over the last six months we have saved and frozen all of the peels and pits from the avocados which we have consumed. A couple of weeks ago, it was time to make room in the freezer and, thus, time to put the avocado remnants to work. The peels and pits were rinsed and dried. Once dry, we ran them each through a food processor. The resulting small chunks have now been "fermenting" in water in large glass jars, one each for pits and peels. Every couple of days we have brought the solution close to a boil to stop any mold from developing. The resulting liquids are currently a lovely, intense shade of red (from the pits) and a more subdued red/brown (from the peels).

This coming week, we will get a couple of skeins of white yarn dyed using each of the two solutions. We also will dye some light gray raw wool. Once the first dying is complete for each set of yarn, we will assess the remaining liquid and perhaps do a second set of skeins in the diluted solutions. The result should be lighter shades of the same colors. Interestingly, the color of the solution does not always translate into the same color that is permanently dyed into the fiber.

*****

I am back. A good piece of time has passed since I began writing about our avocado dying project. We finished our first dye bath and the yarn is now dried and skeined. We were a bit surprised, and a little disappointed, that the solutions from both the pits and the skins did not produce darker yarns. It had been our hope that the large volume of pits and skins would have resulted in more intense shades than they did. In each case, we dyed two skeins (each about 4 ounces in weight) in the bath. When we were finished there seemed to be a significant amount of color remaining in the bath. So, we did a second dye bath. The two sets are both lovely colors. Compared to a similar dyeing we did a year ago, with a lesser volume of avocados, the shades are about the same. We only used alum as a mordant, which generally will result in less intense shades than other mordants. We since have found one source which suggests that a much longer time is necessary in the dye bath for the yarn to fully accept the avocado dyes. We will try this process next time.

Once we had drained the dye solution from the pits and from the skins, we were able to re-grind each group more finely. The re-ground materials again soaked in a water solution for a couple of weeks. If anything, the colors of the solutions seemed to be as intense as the first go-round. We modified our previous procedure. Again, we placed two skeins in each solution, but this time we removed only one skein after the first day and left the second in for a longer period of days. Secondly, we used alum and copper sulfate as our mordants. We tend to shy away from using the other, more toxic mordants; it is unfortunate that there are not a greater choice of environmentally and health friendly mordants!

The second dye bath produced mixed results. We had been disappointed with the intensity of the colors with the first dyeing. The presence of the copper sulfate was immediately evident; both the pit bath and peel bath produced very nice green shades. The skein of yarn in the peel solution was, however, not very consistently dyed. In the end we felt we had better results with our attempts in the previous year, perhaps due to the pits and peels not being frozen. For the moment, we will have to wait until summer for another try; perhaps another guacamole extravaganza awaits us!

Shearing is Over

There seems to be a strange, quiet calm around us currently. Shearing was completed just over a week ago. Now we await the arrival of this year's lambs. So much happened ahead of time in preparation for the two day event that it now seems a blur. In two days, all of our 126 sheep were shorn. Since then, the pace has not slackened. Gretchen is busy with the final skirting of the fleeces. We have begun readying the barn for lambing. In many regards, it was one of the smoothest shearings we have experienced. We did just a few things differently than in the past. I am not sure if those changes made a difference or whether we were just "lucky" this time.

Pre-shearing preparations

In the week that proceeded shearing, we were faced with ominous weather. For us, it has been an extremely snowy winter, accompanied by extended periods of below normal temperatures. The snow and cold seemed to begin within a day or two after we completed breeding in early November. Once we moved each of the breeding groups in from pasture and settled the ewes into the main barn and rams back into bachelor quarters, my first task was to "winterize" the pastures. This job entails taking down and storing all the portable electric fences; grain and mineral feeders need to come back to the barn. We have portable water tanks which are connected by hoses to the water line which runs through all the permanent pastures. The tanks, hoses and connectors all need to be brought home (hopefully drained of water and ice). Lastly, the water lines themselves must be drained. I managed to complete all of the above tasks except putting away the five sets of hoses. All five were laid out down a hillside to drain when their contents thawed. The next day came with snow and the hoses disappeared for the rest of winter. Hopefully they will soon re-appear!

Usually, we hope that late February will present us with a period of warmer weather, which will in turn make shearing seem a bit more comfortable for sheep and shearing crew. This February was not meant to fit the usual mode. Snow storms followed snow storms and the temperatures dropped further below normal. Consistent overnight lows were just above 0°F (-18°C); the highs struggled to reach above 10F° (-12°C). It was not ideal weather for shearing. Dave, our shearer, had given us a date to begin shearing. Since his schedule only gets busier, we had to fit shearing into the date rather than wait for warmer weather. Unfortunately, the prospects for warmer temperatures after shearing were also poor. With the dates for shearing set, we were at least able to line up the crew that we needed to back-up the shearer and make sure things went as smoothly as possible. Besides the two of us, we usually figure that we need at least three other able bodies to help us. At least our volunteer help managed to arrive between storms, a couple of days prior to shearing scheduled for Friday. (Helen and Russ came from Atlanta, Georgia and MJ from Madison, Wisconsin). Thursday afternoon Dave pulled in from his previous shearing job in east central Wisconsin; the next big blow and snow storm was literally right on his tail. No sooner had he gotten his equipment unloaded and into the barn, than it started to snow. By Friday morning we had 10" of new snow; it was cold; a bitter north wind was blowing. We managed to get the barn shut down as tightly as possible. The sheep, of course, were still quite comfortable with their full fleeces. Their warm bodies managed to keep the lower level of the barn in the mid 30°F (-1°C) range overnight without it becoming too humid. Those temperatures are still not ideal for shearing as the heavy lanolin content in Corriedale wool tends not to soften up as much as one would like. The wool is thus more difficult for the blades of the shears to cut, slowing the shearing down and resulting in more work for the shearer.

We made one significant departure from our shearing schedule of years past. Normally, we give our CD&T booster shots to all the flock as we catch each sheep to turn over to the shearer. Combining shots and shearing is a labor saver and usually means that our timing of the booster shots is pretty good for maximizing the immunity that the ewes will pass to their lambs. This year, since we were not certain that all our help would be here due to weather, the two of us opted to give all our shots a couple of days prior to shearing. It meant crowding the flock into an area in which we could easily catch them. The entire job was one that Gretchen and I could handle in less than half a day on our own. Giving shots ahead of time also permitted us to sort out our replacement ewe lambs from the rest of the flock and get them relocated into the newer wing of the barn, next to the ram pen. We do not breed the ewe lambs and they tend to be a distraction during lambing if left in with the pregnant ewes. It makes life easier for the pregnant ewes if their teenagers are not in the same pen. Normally this job of sorting we save for immediately after shearing. Having removed the task of giving shots during shearing freed us up for a smoother and easier handling routine during shearing. The ewes and especially the rams also seemed to be calmer this year during shearing. We speculate that this may have been due to the removal of the extra stress that giving shots might create.

Shearing

Friday and Saturday did not warm up, but we had no choice but to proceed, as Dave had other jobs scheduled. In many ways, our "operation" is a reflection of the same procedures one would see in a commercial shearing shed. The difference is attributable to our fleeces being destined for sale to the handspinning market. Greater care must be devoted to each fleece throughout shearing. We penned the flock tightly near the shearing area so that they were easier to catch. I catch each sheep, making sure that we get the pregnant ewes sheared first, in order to minimize their stress. Russ and I move each sheep to the shearing area where we remove their coats. (The coats are neatly set aside in an area where they will not be further contaminated. This process makes it easier to begin the coat washing marathon which will parallel our post-shearing wool processing.) Once the sheep is undressed, we turn it over to Dave, who proceeds to shear each sheep in just a couple of minutes. If the shearing pace was slower (as it was when I used to do the shearing), we would not require as much extra help. As it is, the five of us helpers can just keep up with Dave. The belly wool comes off first and is separated into a "belly" bag by Helen, our "number one" board sweeper. Once the sheep is shorn, the ewe gets to return to the main pen where a hearty breakfast awaits her. (As an aid to the shearer, the sheep have not been fed since the previous morning.) As soon as the sheep is off the shearing board, Gretchen and MJ scoop up the fleece and place it on a skirting table where a quick, preliminary skirting takes place.

271

The dirty edges (especially around the unjacketed butt) and feed contaminated wool around the head are removed. The neck wool tends to collect hay chaff during the winter feeding. If it is not removed at this point and, instead, gets rolled up with the rest of the fleece, it will contaminate the otherwise squeaky clean fleece. When the fleece has been skirted, it gets placed in a separate bag along with a tag that identifies the "producer". The nicer wool from around the neck area, which is not covered by the sheep's coat and which is contaminated with some vegetative matter, gets placed in separate bags sorted by shade. Eventually we will further sort these skirtings into color blends which we send to Blackberry Ridge Woolen Mill, in Mount Horeb, Wisconsin. There it will be washed, cleaned, carded and spun into yarn for us to either use or sell. Making yarn proves to be a good use for wool which otherwise should not be offered for sale to the handspinning market. Simultaneous with the skirting, Helen is sweeping up the small cuts of wool left in the shearing area so that they do not contaminate the next fleece as it is sheared. She also cleans the area around the skirting table for the same reason. These sweepings also end up in the "belly bags". As soon as Dave's shearing area is clean, we have another sheep ready for him to shear and the procedure begins again.

On Friday, we managed to get all of the presumably pregnant ewes sheared, plus a few of the "old ladies". The shearing total was 80 for the day. On Saturday, we began the routine a little later than the previous day. We had 46 sheep to shear that day: the rest of the older "open" ewes, last year's lambs and finally, the eight rams. Needless to say, we all were tired after shearing was over. Gretchen had prepared a bounty of food for all of us. It was well consumed. It is a busy two days, full of hard work. However, it is a good time for all of us to share with each other. The comradeship makes the two days especially rewarding.

Post-shearing

Saturday afternoon, after shearing was completed, I was able to clear a narrow path through the snow for Dave's van. On Sunday morning, he was able to back through the narrow corridor out to the road and then head off to another job in the area that day. The rest of us got the barn re-organized and as comfortable as possible for all of the naked sheep. They were all cold, but not excessively so. We were thankful that all of them were coping quite well with the conditions. Every one of them continued to eat well. In fact their hay consumption is up, as expected: more fuel for 126 sheep furnaces. In reality that is one of the reasons we shear immediately before lambing: increase the ewe's nutritional intake at the point where the unborn lambs are growing most rapidly. During the next couple of days, the rest of our crew departed for their homes; again everyone managed to avoid any bad weather. Without their help, we would have never survived!

The next phase is just beginning. Each day we bring seven fleeces into our heated basement, where they are spread out on skirting tables to air dry. Here, under good light, we can perform a final skirting on each fleece. We are then able to weigh and evaluate each fleece and prepare it for sale. When all the fleeces have been processed, we will then sort them by colors/shades or special patterns. Then we can compare each to similar types. It helps us decide on prices. It also allows us to select the fleeces we will keep for our own handspinning.

Seven fleeces are the maximum for which we have space at one time. Simple mathematics will tell us that under the best of conditions we will have all the fleeces finished after 18 days. "Simple mathematics" does not consider that it is only 12 days between the last day of shearing and the first scheduled day of lambing, nor does it consider that somewhere in that period other things will come up for which we have not planned. If life proceeds "normally", we will have a couple of lambs born a day or two early. In any case, the fleeces definitely will not be ready for sale until lambing is over and we have had a moment to catch our breaths.

We have also begun the lengthy task of washing (and sometimes repairing) all of the sheep coats we recently removed. Within two days of shearing, the rams and all the ewe lambs have been fitted into smaller (and clean) coats. These two groups have now been moved back into the barn addition which will keep them out of the way and mischief during lambing. It tends to be a little cooler in the addition, so the new coats also serve to insulate them a bit more. At chore time, during the week we are also re-coating the open ewes. The pregnant ewes will not be coated again until they have delivered their lambs. At least by mid week after shearing, the weather moderated. It is warm enough to melt lots of snow and to open up the barn for better ventilation without sacrificing ewe comfort. Now if it would just stay that way throughout lambing....

Big Barn Update

By Nanoo Nanoo,
editor for Baa Baa Doo Press

I see that the folks up at the "house" are telling everyone that they have been *ever so* busy, what with dyeing, shearing and the like. I guess that it is time for me to set things straight again. I will admit that for a day or so those shepherd folk were scurrying around the barn a lot, making sure that we got properly sheared. You would think with all the help they had that they would not have gotten so tired. Speaking at least for myself, Nanoo Nanoo, it was high time that we got rid of all that wool. After all, the real business is at hand. I am pregnant, as are most of my buddies and those babies are on the way! Now we will really get to see that old, bearded, shepherd guy and the nice lady hustle their buns off!

Spring (she is one of the flock, not Mother Nature's offspring) at least got the shepherds attention Wednesday night. She decided that she was not waiting for her due date. After a hearty helping of grain as part of her dinner, she decided to deliver her lamb. Everything went well and we now have the first new member of our group for the year, a strapping little eleven pound guy they are calling "Winter".

That reminds me that I should explain what those shepherds do to come up with names for us. (We have our own names but it is in an alphabet and language too complex and difficult for a shepherd to understand.) Each year they use a different letter of their alphabet for all the lamb names. This year they are using "W". Last year was "V". As might be suspected, I was born in the "N" year. I am told it helps the shepherds keep track of how old we are. As you can see, they need all the help they can get, especially the old breaded guy! Then they do something dumb. A couple of years ago they purchased some sheep from another flock and they keep using the names with which those gals came. Lady Harriet Vane was not born in the "L", "H" or "V" years, nor was her mother, Lucy. I think that you get the picture....

Portia, Opal, Tess and Rubina

Before I could even finish writing these few paragraphs, a bunch of the girls got into the act. This morning Opal, Portia, Tess and Rubina all delivered lambs. The day is hardly half over and I am already behind. Somehow I have to figure out how to type faster. With just two toes to a foot, typing is a challenge. I have the sinking feeling that I am not going to be able to keep up with all the labor and deliveries!

One thing was settled during shearing that is relevant to my literary efforts. We now have a reliable source for pictures of what we are doing. One of the people who came to help with shearing was the old shepherd guy's brother. He is actually a nice chap with a nice laugh. Among other things, he is taller than the old shepherd guy and he is a photographer. His wife also came to help. I remember her because she took really nice care of all of us a few years ago, when the shepherds took off to visit New Zealand and

276

the ancestral homelands of all of our Corriedale tribe. One of our group, Queso, got to talking about cameras with this brother. She arranged with him to borrow a nice, ovine friendly digital camera.

Queso is known to many visitors to the Bed & Breakfast, as she is usually one of the official "greeters" when the shepherds bring the guests out to visit us. Since Queso has decided not to have lambs any longer, she has volunteered to be the official flock photographer (appropriate for a sheep named "Cheese" in Spanish). Now I should have some good pictures to supplement my written efforts. All I need to do is to figure out how to publish them!

I am not due yet for a couple of weeks, so I will try to keep all of you up to date as the lambs begin to arrive. I must be off for the moment. Time to stock up on hay before everyone eats all of it....

An Interview with the Nice Lady

Well folks, Nanoo Nanoo never seemed to make it back from dinner the other night. Now she is off somewhere in the barn looking after her new born daughter, Wambam. Prior to going into labor, she asked me to finish off her most recent literary effort. She wanted me to look after things and to make sure that the sheep perspective is not given a short shift while she is on maternity leave. So instead it is me, Queso, the flock photojournalist and, for the moment, special correspondent.

The barn is finally settling down, as all of my pregnant buddies have delivered their lambs. As I mentioned Nanoo Nanoo had a girl. The shepherds claim she is "cute as a button". (Whatever that means, we sheep have no use for buttons.) The girl is growing well at more than ½ pound per day. Nanoo Nanoo is a very good, attentive mother.

Over the last few days I have spent a lot of time with the shepherds, making sure that they do their jobs properly while the ewes are in labor. While we killed time waiting on births, I had a chance to talk with the nice lady about our fleeces. Here is what I managed to get recorded.

Queso(Q): So, nice lady, what can you tell me about our fleeces this year?

Nice Lady (NL): The fleeces this year are really quite beautiful. As you know, I look carefully at each fleece during the days and weeks after shearing. Then I write down all my observations.

Q: What are you looking for?

NL: I look for any vegetative matter and try to pick that out of the wool along with any short little bits of fiber. I test each fleece for soundness, that is, I look for any tender spots in the fiber length that would break as it is being spun. This year we had no fleeces with tender spots.

Q: How long does it take you to look at each fleece?

NL: It depends. The least amount of time is probably about 15 minutes. Some fleeces take up to 45 minutes. Some of your friends here in the barn tend to be hay magnets. Even though you all wear jackets, some of you like to collect vegetative matter along the edges of your jackets. Some of it even works its way underneath the jacket. After I have the fleece examined, I measure the staple length, I try to describe the color and characteristics of the wool, and then I place the fleece in a clear plastic bag with each of your names clearly visible. Finally, I weigh the fleece and record all this information on a chart so that we can put that information on the website in preparation for the annual fleece sale.

Q: Which fleeces have you spun from our flock?

NL: I have spun or am spinning fleeces from current flock members including Justine, Kassia, Luscious, Mindy, Nutbread, Octavia, Prunella, Ruby, Stud Muffin, Tabitha, Trudi and, of course Nanoo Nanoo's and yours. When we first started with the flock back in 1990, I spun a small amount of every fleece from the original 20 flock members. And, I have spun fleeces from flock members that are no longer here.

Q: I have heard that sometimes you add color to our wool. Is that true?

NL: Yes, sometimes I dye the wool. I especially enjoy adding color to the pale gray and vanilla gray fleeces. Those light grays add a nice undertone to the color. I am starting to use more and more natural dye materials. I especially enjoy trying out new plants as possible dye sources. Some experiments work well, others are not so good.

Q: Am I ever glad you do not try to dye our wool before they shear it from us! What happens to the yarn you spin from our wool?

NL: Some of the yarn is used by that grumpy old guy when he weaves blankets, scarves, and other items. I use some of the yarn in knitting things like hats, mittens and socks. I also use some of the yarn as embellishments in felting projects or in dressing the teddy bears I make. Occasionally, I will sell some of my handspun at the Door County Shepherds' Market or through our Art Gallery.

Q: I heard a rumor that each year you make a list of your favorite fleeces for that year. Is it true? And, why do you do it?

NL: Yes, it is true. Each year I pick 5 or 6 of the best (in my opinion) white fleeces and 10 to 12 of the best naturally colored fleeces. I keep my list from year to year to see how my opinion changes and to see which members of the flock are producing consistently good fleeces. Fleeces on this list may end up being priced differently than other fleeces in the flock. For example, this year one of the top fleeces is going to be from Violette, one of Rhoda's daughters. It is a lovely lilac gray lamb fleece, which is going for $15.00 a pound.

Q: Am I on the list?

NL: Oh yes, you are on the list. This year my favorite naturally colored fleeces are from: Limburger, Naomi, Nutbread, Nanoo Nanoo, Queso, Sunflower, Toodles, Tessa, Tallulah, Upsadaisy, Ulayla, and Violette. My favorite white fleeces are: Cynthie, Quiche, Portia, Quazar, Ukiah, and Vanilla. All of these fleeces have lovely color or whiteness, are consistent from front to back, and all have a lovely crimp. It is unusual for the two lamb fleeces (Violette and Vanilla) to be on the list but both are very special this year.

Q: I have heard that each year you usually keep at least a couple of our fleeces for your spinning projects. I cannot imagine what you do with all the rest of our fleeces. After all, there are over 120 of them this year.

NL: Every year we put most of your fleeces up for sale on our web site. People from all over the U.S. and Europe then buy them. In fact, we will be having our next fleece sale this week. You are even welcome to look.

Q: I really am not interested in buying back my fleece. What good would that do me? In any case, thank you for telling me about our fleeces and what you do with our wool. It was also very nice of you to spend so much time scratching me behind the ears. I think I'll go over and see if the old bearded guy will give me a rub for a while....

First Days on Pasture

By Nanoo Nanoo,
editor for Baa Baa Doo Press

It was prior to lambing when I last managed to write, so it has definitely been a while, but I finally made it! I, like so many of my fellow flock mates, have been busy caring for my new lamb. Unbelievably my girl, Wambam, is nearly 60 days old. I thought of writing something sooner, but then we got to go out on pasture for the first time this year. That is always so wonderful and exciting that I just plain forgot that I have been designated the flock scribe and have the sworn duty to present the sheep view fairly and without bias!

Before I attempt to bring everyone up to date with flock life, I would like to thank my good buddy, Queso, for filling in for me while I was busy with my new lamb. She did a yeoman job with her interview with the Nice Lady, but she has begged off further written assignments...she is more into graphics than the written word. Anyway, thanks Queso for filling in for me.

My lamb was born March 26th. She is a ewe; everyone calls her "Wambam", so I guess that is how she will be known. She is adorable (what else can a doting mother say?), and she has already learned that being friendly to the grumpy old shepherd guy can earn her lots of extra points. So I think she has a future with the flock. The shepherds are beginning to fit the chosen lambs with jackets. I am pretty sure that I heard them talking about Wambam getting a jacket when she is two months old. That is usually a good sign that they want her to stay. I am also a grandmother again. Daughter Pookie had twins and daughter Toodles had a single. In addition, I am now a great grandmother, as Pookie's Ulayla had twins.

Nutbread with her lamb, Widget

Today is the 13th day that we have been out on pasture. Has it ever been nice! The chickens got out four days ahead of us. Can you believe that! The old bearded shepherd guy was mumbling something about there being enough grass for birds but not for us. For the first eight days, it was just us ewes with lambs who were allowed out. Then this week, the retired ladies, including my mom, Mindy, joined us. We are still waiting for last year's lambs. The grumpy guy says that there has not been enough grass for all of us. Grudgingly, I have to admit that he is right. It just does not seem to want to rain like it should this time of year. At least it has been cool, so the grass is still growing, but just not as fast as we are accustomed to. Nonetheless, it has been nice just to be out in the sun and fresh air.

The lambs have had a grand time. This year's brunch has been really quite smart learning all about electric fences and how we work our way out to the various pastures. After all, we are already out on the back five acres. That is a quarter of a mile walk from the barn. Except for one day, when Waldemar and Wesley were goofing around in the barn and got left behind, all of the lambs have made it on their own. (Boy, was the old shepherd even grumpier when he had to catch Wesley and carry him all the way out. At least Waldemar figured it out, once they got him out of the barn.)

We have even had visitors out on pasture. The people who stay up at the place they call "the House" have been out to schmooze with us. The Bobolinks also arrived back from Argentina this week. It is always interesting to hear about our Corriedale cousins down on the pampas. So, all in all, life is going well for us. Got to go…there is grass to be eaten!

New Arrivals

We have just finished three days of soft, steady rain. It was a much needed two to three inches of moisture, which seems to have given the spring time flush of forage growth the kick-start it needed. Our apple trees are just about ready to bloom. That event usually coincides with the birth of the local Whitetail Deer fawns. This morning as I was moving the fencing for the sheep to resume grazing in the back pasture, I was "greeted" by one such little creature. He/she was dead-still in the middle of the previous day's pasture. I literally had to step over the fawn to remove the temporary fence. I made sure not to touch him/her to leave any of my scent. I am positive that mom was somewhere in the woods next to the pasture, keeping an eye on both of us. Luckily, she had not left the little guy/gal in the middle of the deeper grass that the sheep would soon be grazing; I would never have seen it under those conditions. The discovery of the fawn by over 200 ewes and lambs would have led to all sorts of chaos.

I returned an hour later to check on both the flock and the fawn. The sheep were eagerly working on their breakfast and the little guy/gal was quietly sleeping in the sun. Just another beautiful spring day....

More Spring Days

It is now the middle of June and we finally seem to have passed the threshold into more full-blown late spring weather. Our apple trees are dropping their blossom petals after retaining them for a wonderfully long period. It has been an extended and beautiful show for the trees. Everything benefited last weekend from a couple of days of steady and significant rain fall. The vegetable garden is almost completely planted. Until this week, the soil temperatures have been too cool for the squash, cucumbers and melons. The cool hardy seedings have finally begun to germinate. The peas, at least, have loved the cool weather. For the peas, we are into the defensive mode of warding off the marauding deer. I have the electric fencing up and the motion sensitive sprinkler is armed and aimed at the pea area. Now all I need do is avoid getting myself sprayed by inadvertently walking in front of it.

Yes, there is the more sinister aspect of the little spotted guy or gal that I mentioned earlier. If he or she makes it to adulthood, it will mean just one more mouth in an excessively large population of deer to over-browse the native trees and flowers, not to mention the plants that we try to grow. I have nearly stepped on a fawn a couple of times since my first sighting. It is impossible to tell if it is the same one. If it is, mom is moving the fawn all over our pastures. It has become much more mobile and more inclined to flee when I accidentally stumble upon it. The next big deer challenge will be to avoid any and all fawns in our hayfield, as I start cutting hay in the coming weeks. Avoiding the turkey nests in the middle of the hay will be more difficult or next to impossible.

The sheep are now grazing in the pasture we have named "The Orchard" (the site of our cherry trees of some years past).

285

Hopefully, they are all quite happy there. It is high ground for our farm, which means that it tends to catch any breezes that blow and make it a bit more tolerable in terms of both temperature and biting insects. Last spring, we "renovated" the forage in "The Orchard" and the results this year are staggering. The renovation consisted of broadcast seeding two types of clover in the pasture just ahead of the sheep. As they grazed, it was hoped that their hooves would help pack the clover seeds into the ground where, with a bit of rain, they would germinate. Last fall it was evident that the seeding had taken. This spring the growth of clover is tremendous and lush.

It is also easy to see where I did not quite overlap with passes of the seed spreader, i.e. narrow strips of grass without any legumes. The clover provides excellent grazing and nutrition for the sheep and also is an excellent natural source of nitrogen fertilizer for the soil. It is too bad that they will finish grazing the area just as the clover was about to be in full bloom. The smells would have been sweet and the sight almost as breathtaking!

Mulleins

Gretchen has been busy with a bunch of natural dyeing projects. Earlier this month we picked a couple of buckets of Dandelion blossom for dyeing. We sometimes get the giggles when we think of the sheep watching us, on our hands and knees picking buckets of dandelion blossoms. One can almost hear the combination of indignation and consternation in their mutterings. "Why don't they leave those blossoms for us? They're the tastiest part! Well, we already knew they were a bit daft, but this beats the cake!"

The most recent dye project used the leaves from Common Mullein. It is an alien weed with large, flannel textured leaves.

286

Later in the year, it will produce tall yellow spikes of flowers. It seems to like thin, poor quality soils, which means that it thrives in the eastern edge of our Pasture #4 (a.k.a. "The Rock garden"). For whatever reason, it is one of the plants that the sheep refuse to graze. As such, the sheep are of no help in eliminating it from the pastures. Rather than just pulling up the offending plants and casting them on a compost heap, we cooked down some of the leaves into a dye solution. They yielded a number of lovely shades of yellows and browns. If we waited until later in the season when the mullein blooms, the tall stalk of yellow flowers produces a beautiful yellow dye.

Eventually the resulting dyed wool and yarn will work their way into our spinning, knitting or weaving projects. If you happen to visit the Gallery this summer, you may see the big, red crock-pot bubbling away on the back porch. It is getting a good workout this spring. Just ask to see "what's cooking"; it can be quite interesting.

Respecting Fences

This spring I have had correspondence with someone brand new to sheep and shepherding. Like many of us when we began our flocks, my acquaintance is brimming with eagerness and enthusiasm for the adventure to begin. Each of us probably approached starting a flock differently than the next shepherd. As a result, our initial experiences can be vastly different. Rather than waiting to purchase spring lambs when they are weaned, my friend thought that raising a couple of bummer lambs (i.e. orphan or abandoned lambs) would at least help her family get their feet wet before they added more lambs in summer.

Raising abandoned lambs can be an extremely rewarding experience. It can also be extremely exhausting and frustrating. Over our time as shepherds, we have tended to become less enamored with the prospect of raising a bummer. When we are faced with the absolute necessity, we try to make sure that the lamb(s) grow up thinking that they are sheep rather than human. We were not always like that. Our first bummer, Hope, a true orphan, spent her first week or two of life in a large cardboard box in our house. Transitioning to life in the barn was difficult for her and for us. Nonetheless we all survived: at 13 years of age Hope is now the oldest sheep in the flock. She proved to be a wonderful mother, is still our good friend, and often seems to think that she is more "human" than the rest of the flock. Our subsequent bummer lambs have spent no time in the house and have been part of the flock from the beginning, even if they were in a separate pen for a time. None of them have been as "humanized" as Hope was. It has made for fewer problems for us and the sheep.

From her description, my friend has been both rewarded and frustrated by her experience raising a couple of bummers. It sounds as if the two lambs she started with have had much more human contact than our Hope ever had. In addition, they have had no "regular" sheep contact. As a result, when it was time for them to "become" sheep, they revolted. To be put in a pasture and be expected to contentedly graze without any mentoring was a concept entirely foreign to them. The fence that separated them from their shepherd was a barrier that had to be surmounted, circumvented and overcome. At last report, the fence was loosing the battle and the new shepherd was finding the experience less enjoyable than initially imagined. I have not seen the sheep in question nor have I seen the fencing that they are battling. Therefore I cannot make any direct observations. I can, however, at least relate to the problem and offer some thoughts.

Good sheep fencing is both a physical and psychological barrier. Different breeds of sheep have different ideas about fencing; some are easier to contain than others. In different situations, the blending of physical and psychological barrier must vary in portion to each other. In our particular case, we have chosen a breed (Corriedale) that is relatively mild mannered about a fence. It is not a breed that tends to have individuals that wish to go off on their own. If one individual feels so inclined, their need to stay with the rest of the flock is generally overpowering. We have had very few "bunch quitters". (It is a term I picked up from a Nebraska cattle rancher who seemed to specialize in the "Breed".) For us a "bunch quitter" is a sheep that decides that grazing in the adjacent pasture has to be better than where everyone else is grazing. Inevitably, it is a behavior that begins in a lamb who learns that it can drop its ears, keep its moist nose down and let its wool insulate it from any electric shock from the portable electric fence. Once the lamb has mastered the technique, it becomes a "bunch quitter" and it is a behavior very difficult to change. When we have had a lamb that insists on becoming a "bunch quitter", it is a guarantee that the lamb is one of the first on the trailer to the sale barn. If there is a genetic component to "bunch quitters", we have been able to select against it.

For many years now, we have followed a pretty standard training procedure when it comes to introducing our lambs to electric fencing. On the first day that the lambs get to leave to barn in early spring, I make sure that our permanent perimeter electric fences are operating at peak performance. There must be no shorts. This condition also applies to the movable fencing which attaches each day to the permanent electric fencing. I also string three strands of movable fencing across the pasture, the lowest of which is at least a half inch wide electric ribbon, rather than the smaller, less visible but much more easily moved narrow cable. The first day's pasture is relatively small compared to the regular space the flock will have in subsequent days. When the fences are all set, we let the ewes and their lambs out onto the pasture. Curiosity is usually the first cause for an electric shock for the lambs. The white electric ribbon is especially intriguing if it is fluttering in the wind. It is excruciating watching each lamb make its acquaintance with the fence. You know that there is going to be contact, a shock and then the resulting race to find mom. You dare not warn them even if you could. We are now to the point where, after one day, the vast majority of the lambs will have learned to respect the fences and after two days it is an absolute majority. I am now convinced more than ever that the older ewes are our best trainers when it comes to lambs learning about fences. Somehow they are communicating some message to their lambs to stay away from or at least respect the fence. Years ago the training was not nearly as efficient and effective. Every year it seems to improve. The equipment has not changed nor has our approach. In some fashion, the ewes are training the lambs better.

This year two lambs convinced me the training was indeed working well. When we started grazing this spring we worked our way rapidly through the first few pastures so as not to overgraze the still relatively slow growing forage. By the ninth day of grazing, the flock had already been through two pastures and was starting the third. At the most, the lambs had only been exposed to portions of each of the first two pastures (and never the same portion) for four days. They were now moving down long, narrow electrified raceways through each pasture, making from one to four sharp turns without a mishap. The morning that we started the third pasture (about a quarter of a mile from the barn), two of the larger lambs got left behind goofing around in the barn. By the time I finally convinced the two lambs to exit the barn, the entire flock was out of sight. The two little guys were frantic, but we eventually got them going down the raceway through the first pasture. They made the first left turn and then the next right, all the while not touching the fences or trying to go through them. We got to the third pasture which was new to them and things became a lot more panicky. Before the rest of the flock would be in view, they had to pass through a gap in the old stone fence and then make a final left turn. One of them made it and was back with mom in an instant. The second little guy finally forgot about electric fences and balked at following his buddy through the gap in the stone fence. Instead, he back-tracked, panicked and crashed through the two strands of portable electric fence. He was now in the second pasture which was entirely un-subdivided except for the raceway. He headed to all four corners, many times, but never once tested any of the perimeter fences. We could not catch him, nor could we convince him to go under the portable fence if we lifted it up for him. Eventually, we took down part of the portable fence, making a gate for him. He passed through and then retreated all the way back to the barn! There I was finally able to catch him. My only recourse now was to throw him over my shoulders and carry him the quarter mile back to the new pasture. Luckily he was the lighter of the two!

What the entire frustrating experience did tell me though was that 1) the two lambs already were well schooled in the electric shock they expected if they challenge the fence and 2) they had already familiarized themselves with the ins and outs of about ten acres of pasture and could definitely recognize when they came to new territory. For lambs not yet two months old, they had definitely already stored away a lot of memory data. (By the way, they also have never again let themselves be left behind in the barn!)

The flock has now grazed through the third pasture and two more smaller ones. As of today, they have been in the pasture we know as "The Orchard" for three days. Their navigation skills are still sharp. The adult ewes seem to know where they are going even after not being there since last fall. Last spring, we decided to renovate the orchard pasture by over-seeding it with two types of clover immediately prior to turning the sheep into the pasture. The hoof action on the moist, spring ground followed by further timely rains resulted in a nice growth of clover by fall even with repeated grazings. This spring the clover growth has been phenomenal. The sheep are in heaven. Yet despite the lush growth all around they have chosen not to challenge the little two strands of electric fence which separates them from today's and tomorrow's grazing. It is a good testimony to their fence training and their respect. It is nice to see that we actually have come a long ways from our early days.

Early morning grazing, on spring pasture

Making Hay

June is now history. In the space of a month, we have gone from spring (a frost in the first week), to full blast summer (temperatures in the upper 80° F (30° C), to cool and damp. If one farms in Door County, June is the time for cutting and baling first crop hay. For the large farms or farms with an abundance of available labor and/or equipment, the job is usually over in a week or two. For us, it would be nice to have the entire job done by the end of the month, but that rarely occurs. This year is no exception. The trick for making hay is all in timing. The hay must be cut when it is not wet and when there is a forecast for a couple of days for warm, sunny, drying weather. If it all works out, we can get the hay into bales which are put away in the barn before the next rain. If rain sneaks up on us earlier than expected, we are sometimes able to get the hay baled and then hurry to squeeze the loaded wagons into the machine shed. Later after the rain passes, we can roll the wagons back up to the barn and unload them into the mow. Since 40 acres of hay results in more than four wagon's worth of hay, this entire scenario is re-played many times before it is done.

While we would like to depend on the forecast to be accurate, it often is not. It is more likely that the forecast will change as soon as we have cut hay, but before it has had time to thoroughly dry and cure. So far, we have been both lucky and not so lucky in that regard. Once it warmed up, we managed to get a good start on cutting the hay. The first cutting around the 40 acres is obviously also the longest. Our haybine cuts a nine foot wide swath. Each subsequent round of the field becomes a slightly smaller concentric image of the previous round. I have now cut the field often enough that I know that in total there will be about 62 rounds.

Our hay operation is in many ways antiquated, in that we cannot rely on all the efficiencies of modern agriculture. If we had a modern barn facility, we could bale our hay in large squares or round bales. Such bales, weighing up to 900 pounds, require only one person to operate the baler and later, one person to collect, move and stack the bales mechanically with a fork attached to a tractor. As it is, our barn was built before the days of any type of bale. We cannot get large bales into our barn, but at least the barn is able to handle the small bales of 40 to 50 pounds. Even if we could get the large bales stacked into the hay mow, it would still be impossible to feed them to the sheep in the barn. The space for the sheep is on the lower level and that level is too confined to allow a large bale to be moved inside.

Without a major barn rebuilding, we must continue to bale our hay in small bales. Small balers can be set up to self-load onto/into a wagon, but there is a labor trade off here. The self-loaded wagons are not stacked with bales but rather receive indiscriminately thrown bales. They must be unloaded manually and, for us at least, they tend to be more work than they justify. To be efficient, small bales require a second person to stack the hay as it comes off the baler onto the hay wagon. There are just the two of us and by afternoon, when the hay is dry enough to bale, Gretchen is manning the Art Gallery. That generally leaves the stacking to me, and I am also the driver. If I work it properly, I can slide up to 10 bales onto the wagon before I need to stop, hop off the tractor and onto the wagon to stack the bales. As the wagon becomes fuller, the stops become more frequent. It eventually gets finished, but not as fast as it would with dependable help.

Times have changed, however, even for us. When we first moved to our farm and before our sheep arrived, I worked on a local dairy farm. I did pretty much all the tasks that the "hired hand" would do, from milking cows to helping to bale hay. My educational curve at that time was steep, but at least I was exposed to most of the tasks necessary for the successful operation of a dairy farm. It was in this capacity that I learned the intricacies of making hay with equipment that was up to date and to a scale large enough to handle the task. In contrast, the equipment that we owned on our own farm was primitive by any standard of the day. We owned a moderate sized tractor, but no haybine for cutting. We also owned an old, small International Harvester Cub tractor which came with a four foot wide cycle bar cutter. Cutting hay with "Little Oscar" was slow and tedious. Because the hay was cut but not conditioned, it also dried slowly.

Nonetheless, it is still a fond memory of cutting my own first field of hay (albeit quiet small) and actually getting it baled. It is a memory sweetened by the fact that it was one of the few times that my father was able to visit and help with a task he fondly remembered from his youth. Our baler, at the time, was as old as the little tractor. It had seen lots of service over the years on this farm. I was eventually to learn that many of the local farmers loaded hay behind that same baler when they were kids. At the time, the baler was shared amongst many local farms, much like lots of the other equipment and tasks.

I quickly saw the need to invest in a haybine (a cutter that also conditions the cut hay). We purchased a used, seven foot wide New Holland. It ran off of our larger Massey Ferguson tractor ("Big

Oscar"). Hay cutting efficiency improved. Eventually, we decided that we had "outgrown" the little seven footer. We purchased a nine foot wide model; with each round cut we had added an additional two foot swath of hay and, therefore, sped up the process significantly. The old baler was eventually replaced by a "newer", old model, which had spent all it life on the farm of an old Norwegian. Eventually, it too was replaced by a newer model which runs a bit faster, has a wider pick-up mouth and is a lot more reliable.

Regardless of which baler we used, after all the wagons are full, they then need to be unloaded. That is again a two person job (one on the wagon to unload onto the elevator and the other in the hay mow to stack the bales as they come off of the elevator). Here at least, the two of us can work together. Since the barn is next to the Gallery, Gretchen can unload as long as we do not have Gallery customers.

This year we still have over 600 bales left from last year, which is always a nice cushion should we come up short with our current crop. When and if each of the three mows are full, they will hold over 3000 bales. As I write, we have baled about 550 bales and they are all safely stacked in the first mow. Obviously, we still have a long way to go before we will be comfortable with our supply of hay for next winter.

During this last week, the operation ground to a halt. It turned cold and cloudy. Scattered showers became the norm for the last few days; often the rain was followed by periods of mist or fog. It was not weather that permitted any further cutting. Unfortunately, we guessed wrong before the rain began: the result is five rounds of hay that were cut the day before the moisture arrived. It still sits on the ground awaiting sunshine and warmth to dry. The moisture does the cut hay no good nutritionally, as it lays there on the ground.

Presumable the warm, dry weather will return. When it does, we will have to get back to work.

Hard Working Sheep

By Nanoo Nanoo,
editor for Baa Baa Doo Press

I t is high time that I let the world know what has ***really*** been going on with the flock. The old bearded shepherd guy would have everyone think that he has been ever so busy, what with baling hay and such. But after all, it is what he is supposed to do! There is no need to give the guy any slack. (If you have not already surmised, this is Nanoo Nanoo writing. The world does need the ovine perspective!)

We have been quite busy, which is why I have not had time to contribute more to the book. What the old bearded guy forgets to tell everyone is that we, the sheep, have been harvesting hay nearly everyday since mid-May and it is now Mid-July. That is a lot more work than he has accomplished. We have now been through just about all of our regular pastures twice. We seem to be eating through the pastures faster of late. It has not rained very much for the last four weeks. (I guess that we cannot blame that on the old guy.) As a result, the pastures are just not growing anymore. We do not have as much to eat in the same area as earlier, so the old shepherd guy gives us larger areas to graze every day. I certainly do hope that we do not have to start eating that baled hay while we are on pasture (like we did in the last drought of a couple of years ago).

299

It just does not compare to fresh green grass and clover. The old shepherd guy is not letting the rams out on pasture anymore. He says there is not enough for them. The same is true for the ram lambs. They took those eleven boys away from us as they were behaving a bit too much like their fathers. They are also relegated to eat baled hay in the barn. Rumor has it that they are still getting grain, unlike the rest of us.

In addition to harvesting forage, we are keeping up with our job of educating and entertaining the guests from the bed and breakfast. It is a difficult task, but we are up to it. We just had a bunch visit. They came out with the nice lady and the old guy. There was a couple from Milwaukee who we remember from previous visits and some new people from Arkansas (not that we really know where any of those places are exactly). As usual we had to show off the wool under our jackets and then let them pet us. We even are getting the lambs trained for duty. Today little Wizzbang, Tabitha's daughter, was learning the ropes. She really has become quite the little "suck-up".

I think that just about all the lambs that are getting jackets now have them. (Wizzbang has hers.) The shepherds have not given out any more of them recently. By my count there are now 28 lambs wearing coats. That is good; that means they all get to stay here! A lot of us adults are getting our jackets replaced. It is about time too, since we have been growing lots of wool and those of us who had lambs are starting to fill out a bit after producing all that milk.

300

Wendolyn, showing off her first jacket

It is also about time for the lambs to learn that they cannot have a drink from us whenever they want it! Someone needs to tell Puss Puss to stop feeding those twins of hers so much; she is just running out of energy. In any case, it is nice to get roomier jackets, especially since they are cleaner than the old ones which now are soaked in lanolin. It is nice that someone is recognizing how hard at work we are!

Sensommer

The Scandinavians have a term that describes the transitional period in late summer just before the arrival of fall. In Danish, it is "sensommer". Like so many words in different languages, a literal translation does not always capture the true feeling of the word. Thus, it is that "sensommer" conveys something more than merely "late summer". Quite often it seems to be a time of year that we only briefly experience in the upper Midwest. This year seems different. At the moment, we are in a period which more closely resembles the Scandinavia season. Days are cooler than normal and alternately blessed with sun or rain. Nights are cool, but not uncomfortably so. Fall is not here yet, but one senses that it could either arrive soon or dally around a long while before making an appearance. In many ways, I truly sense that we are experiencing sensommer.

The Rains Finally Fell

Not that long ago I would never have believed that we would reach the season in which we now find ourselves. It is amazing the difference that a couple of inches of rain can make. It renews the soil and hence the plant life. It cleans up a dusty world. It brightens the soul and the humor, both for humans and sheep.

Over much of June and all of July, we received **very** small doses of rain. The result had been a virtual stoppage of any pasture re-growth, a reduced volume of first crop hay and no second cutting of hay. The vegetable garden seemed to refuse to grow, despite liberal amounts of irrigation. Our humor started to flag. Perhaps the only factor that seemed to soften the dryness was that it has also been unseasonably cool. The most dramatic effect was the loss of any grazable pastures for the sheep. In early August, we were forced to import a semitrailer load of large square bales of nice hay. The day after the hay's arrival we began feeding it to the sheep. Each bale weighs between 800 and 900 pounds. To provide enough space (barely enough) for the flock to all get at the bales requires feeding four bales at a time, with each bale surrounded by its own expandable feeder. The older adults had experienced this routine before and recognized immediately what was happening. We, therefore, had little need to teach the lambs and younger ewes what was going on. They just followed their elders' lead. Within five days, the first four bales were gone. After setting out the next four bales, we were settled into what looked like a long and less than enjoyable experience. With a total of 51 bales, we figured to be able to survive for a couple of months. If there was no rain in that time, we would already be taking more drastic steps to survive.

Finally, in early August, we received over an inch of rain and, in the next week, we enjoyed more than an inch and one half. The new moisture managed to get the pastures to begin greening up, but it will still be a long while before they are again deep and lush. Hopefully, we have turned the corner.

The vegetable garden produced the most dramatic response. Tomatoes and peppers finally looked healthy and began to produce. Squash, melons and cucumbers that had all refused to grow now seem to march across yards of the garden in a day. It is, once again dangerous to walk too near the zucchini; one may get hit by a fast growing green monster!

The weather patterns this summer have produced strange harvest schedules. The cherry crop was a couple of weeks behind and the fresh pea harvest a month late. It was strange to be picking peas at the same time the cherries were ready for harvest. At least this year's cherry crop is strong, compared to the virtual non-existent crop in 2008. Most times I do not miss the work we had when growing our own cherry orchard. However, the first day or so of harvest is still exciting, even if it means picking from someone else's orchard.

Lastly, not to be out done by anyone, our colony of barn swallows is busily feeding their second brood of chicks for the year. The earliest of the new broods has just launched and it appears that the rest will be ready in the next week or two. It is a special treat to have the adults and the earlier chicks all flying over the pastures in the early morning, vocalizing their excitement for the day to come. Like the rest of us they seem to have a special appreciation for the recent rains.

Buckthorn

Gretchen's dye pot has been busy of late, taking advantage of some of the late summer flowers, berries and mushrooms. As I write, the first batch of yarn has just been removed from a dye-bath of Buckthorn berries. Buckthorn is an alien, invasive scrub which grows to tree size. Its presence is definitely not appreciated and we make a concerted effort to limit its spread. But we can have the best of both worlds: use the berries as much as possible and still try to eradicate the plants. Having said this, I must point out that its berries make a wonderful dye. The resulting color is usually a deep

304

avocado green shading to bronze, which is extremely surprising since the berries appear quite purple at picking.

On the future dyeing agenda is a large mushroom which rapidly appeared in the front lawn after our first heavy rains of the summer. Dye color is unknown; it may be a total waste. The wild Goldenrod is starting to bloom in full force. It is always dependable for a good dyeing session or two. We are also in the midst of dyeing with leaves from a couple of species of eucalyptus tree. No, we are not able to grow eucalyptus in our climate. However, one of our good wool customers from California provided us with a collection of dried leaves from two different types of trees. Since the leaves are dried, we have been able to dye with them whenever the time was suitable. So far, they have yielded three different shades of orange, the last batch being the most intense.

As if wanting to add to the late season color festival, the Monarch butterflies are on the move. It seems that it has been a good year for Monarchs, based at least upon our casual observation. Unfortunately, the same cannot be said for most of our usual species of butterfly. Their absence has been significant this summer. The last Monarch generation of the year is now emerging and intent upon stocking up on energy as they begin their flights to Mexico. The Purple Coneflowers in our garden have been a special favorite for them this year. Here's wishing them a safe journey to the south.

Of late, we have been able to visually experience "reds" that one can never expect to reproduce with a natural dye. Last week, as I was attempting to bale a very late second cutting of hay, I noticed what appeared, from a distance, to be a red survey stake in the 20 acres of woods just west of our large hay field. I thought it strange that someone would be placing a stake in the woods because: 1) it is our land and no one had contacted me about it, and 2) the woods are a wetland, primarily a semi-open ash and cedar swamp much of the year. I promised that once the hay was baled I would return to investigate this intrusion into our domain. When I returned, I discovered that the "marking stake" was actually one of many Cardinal Flowers (*Lobelai cardinalis*) in bloom. For our area, it is a relatively rare flower which blooms in late summer usually along the edge of dry seasonal creeks, one of which borders our woods. I usually see their blooms along the creek, but this was the first time I have seen them in the middle of our woods. Once I slogged my way deeper, into the woods I discovered that the flowers appeared to be everywhere where the deepest pool of water stood in early spring. If there ever was a sign of Sensommer, the Cardinal Flowers were it!

Journey to the End of the Earth

By Nanoo Nanoo,
editor for Baa Baa Doo Press

Hello again! It's Nanoo Nanoo here with the ovine perspective. I have been off the grid for a good bit of time. I have not been able to get logged into the old shepherd's computer because of where we have been grazing of late. The only time we could make our wireless connection was in the evening when we were in the barn. We did not dare try that because the old guy might have discovered us using his computer. For a long time, the old grump had us spending our daylight hours in the place they call the Pasture #4. We know it as "the Rock Garden", because there is as much ground occupied by large stones and boulders as there is grass. That pasture is where he feeds us those huge bales of hay when our pastures stop growing in a drought. It is also in a low hollow; Queso tells me that it is a wireless "dead zone", what ever that means! When it finally started to rain in August, our pastures began to grow again. So where does the old guy send us? Way out into the big hay field. It is also in a low spot! Even if it was not in a hollow, Queso and I figure that it is so far from the barn that we would never get our computer connection to work. Only now have we gotten close enough to that other barn where the shepherd guy and the nice lady live that we could get into the computer without him knowing!

The field to which he sent us is where he harvests the hay that we get to eat during the winter months. We usually get to graze there only one time each year, usually in late summer when he does not need to cut it for baled hay. It is really kind of exciting to be there. The Sandhill Cranes often visit us there in the early morning. Occasionally, a coyote will pass by (we just try to keep **very** quiet and still when that happens!). Just yesterday a Canada Goose dropped in and spent the entire day with us. That was really cool! She showed us how to stay out of the way if the bearded guy drives by on his tractor. She just took off and flew over us **and** the electric fence. We ran along to watch but had to stop at the fence. We hoped that she would be there today to teach us how to fly. It would be so much easier to get to the pasture in the morning for breakfast. In the evening, we could just sail back to the barn and forget about all those dumb raceways the old guy builds for us. Unfortunately, our goosie friend was gone this morning so I guess the lessons will have to wait.

Actually, we figure that where we were is really pretty close to the end of the earth (at least in that direction). Just beyond the edge of the field is a thick line of trees that you cannot see past. None of us have ever been that far, so we are pretty sure that no one can go much farther. My grand daughter Wascal and her buddy Wallflower left the flock while we were out on the extreme edge of the hay field. The nice lady told us they went to a new home in a place called Eagle and that they were going to be very happy there. We figure that Eagle must be in a different direction, probably off beyond the swamp where the Cardinal flowers grow.

Grazing near the end of the Earth

All this talk of going to the edge of the earth reminds me that it is an anniversary this month. The cat called Pussa arrived on the farm four years ago this month. In the flock, we do not usually have much use for cats. The ones that usually show up around here are always so high strung and spooky. They always end up scaring the sheep buttons out of us. Pussa is different; she understands that we do not like flighty, fidgety animals in our barn. So even though she is not ovine, we let her stay. In fact, her first winter here, she spent with us. We let her snuggle up to our warm fleeces in exchange for removing a mouse or two every night.

Pussa has never really said where she came from. The old bearded shepherd guy says that someone just dumped her here, like they seem to do every fall. He may be right, but we have a different theory. The shepherd guy rarely leaves us, but four years ago he returned to Denmark for a couple weeks after many years absence. It was shortly after his return that the cat showed up here. Since she has a common Danish cat name we, with our superior ovine intellect, have deduced that she must have stowed away in the shepherd's luggage. Of course, we are not too sure where exactly Denmark is located. Our best guess is that it is in the opposite direction from the big hay field we were just in and probably beyond Eagle, Wisconsin. In all likelihood, it is just beyond the woods across the road from us. That is where the cranes always seem to fly, so it must be at the opposite end of the earth. We try talking to Pussa about all this, but when we do she just clams up or starts talking in some strange tongue. (We figure it must be Danish!) We do not try to press her too much about this. She is a good friend, but she is definitely not sheep-like. She now has finagled her way into the place they call "the House". That is why she rarely spends the nights with us any longer, but we do not hold it against her. At least, this way we can learn a bit more about what goes on up there, since none of us (except for Hope) has ever been there. Regardless of where she came from, we are glad that Pussa decided to stay here all this time.

Rams and Ewes, Oh My!

As it now seems to be a tradition, I manage to get away from the farm for nearly two weeks every other fall. My time away is spent in Denmark. It is always special to be able to be with my Danish family and my Danish friends. Thanks to all the Danes who make my adopted "home" such a special place. Tak allesammen! It is equally as special to return home to Gretchen and the flock.

As was expected, my return home to Whitefish Bay Farm did not allow for much time to catch my breath. A day after I got back, the last of this year's lambs were shipped to market. We are now down to our "winter" population: this year it is 129. We keep telling people that we are cutting back; last year we over wintered at 127; I guess that we have trouble with our math!

Immediately after selling the lambs, we had to begin preparing the pastures for the fall breeding groups. While I worked on the other pastures, the entire ewe flock was together for one last period of six days in Pasture #3. At least they were spared having to be out there during the heavy rains of the previous two days. They had beautiful fall weather to be together as a group. The ash, birch and maple were starting to show nice fall colors and there was still plenty of grazing left in #3.

As September and October tend to be one of the most attractive times of the year to visit Door County due to the fall colors, we also have been exceedingly busy in the Art Gallery and B&B. Gretchen, especially, has spent hours in the Gallery. Yarn and finished fiber products have sold unexpectedly well this year, for which we are always glad. The pace has, however, been frantic at times. It will be very nice, in a couple of days, to finally close the Gallery for the season. Gretchen finished up spinning fleeces from Nutbread and Ruby while she manned the Gallery. In addition, we continued with our dye project with Cosmos flowers from the garden. It is truly amazing, the lovely yellow color they produce, considering the intense red and violet shades of their flowers.

In the midst of all the other activity, we devoted an entire day (as we always do in October) to sorting the ewes into breeding groups and getting them together with a ram, each in their own separate pasture. This year we decided to cut back on our breeding numbers. There are 63 ewes currently with four rams (that is a drop of over 20 ewes from last year). While we plan to retain about the same size adult flock next year, there will be more ewes who are "retired" and whose sole major job will be the production of wool. We are scaling back significantly our emphasis on selling breeding stock. As of October 13th, Rhett, Ulmer, Stud Muffin and Vermicelli were each giving their own group of ewes. Breeding appears to be going well. With less than a week gone, over 50% of the ewes have already been well marked, despite often miserable rainy, cold weather.

This year we also decided to try to keep the breeding groups as close to home as possible. It is the only time of the year when the sheep (at least the breeding flock) do not come into the barn at night. It is physically impossible for us to keep all of the groups separate and still be able to bring them inside each evening.

312

Throughout much of the summer and into fall, we have had what sounds like a pretty good size pack of coyotes in the neighborhood. Their howling has been intense most nights. While we have yet to experience any loses due to coyotes, we do not wish to loose any sheep to them. Hence, our breeding groups were all as close to home as possible. They were hopefully protected by as many layers of electric fence as we can possibly set up for them. We are never really comfortable until we declare breeding over and everyone returns home.

The Cycle Has Begun Again

The breeding season officially ended as of November 19th with all of the flock back in the barn and the rams back in bachelor quarters. The end occurred none too soon, as the deer hunting season begins on the 21st, and it is thus prudent to make sure that sheep and shepherds are out of the line of fire. Those familiar with deer hunting in Wisconsin know what a big event those next nine days are. The weather for breeding was generally miserable: cold and rainy. We did not have any hard freezes and, therefore, our pasture water lines did not freeze-up for the year. Breeding lasted about four days longer than planned, due entirely to a severely sprained shepherd's ankle. The extra extension did not, however, result in any further breedings. How do we know? What follows is an abbreviated lesson in controlled sheep breeding.

There is a wide extreme in terms of how shepherds approach breeding their flocks. Outwardly the easiest method is to put one or more rams in with the ewes for a given time, usually at least a month. When the rams are removed, the hope is that the ewes are all bred. Short of later use of ultrasound, the only way one can measure success in this system is to wait 148 days and see if lambs begin to appear.

The other extreme is to place a ram with a given group of ewes (a good ram should be able to deal with at least 50 ewes) and closely monitor their performance. This monitoring is done by placing a harness which holds a large crayon on each ram. When the ram mounts a ewe, he leaves his mark. The observant shepherd makes note of the date. When the ram has been with the ewes for about 17 days, the crayon is switched to a different color. Seventeen days is the usual heat cycle for sheep. If the ewe has not become pregnant, she will come into heat again in 17 days and the ram will again try to breed her and thus leave a new color on her rump. If he does not remark her, it is a good sign that she is pregnant. The marking harness is the system which we employ. It is much more labor intensive for us, but we feel that the advantages outweigh the extra work.

1. We have a much better idea if each ewe is pregnant. (However, there is no guarantee that she will carry her pregnancy to full term.)
2. We know exactly when she was bred and therefore when she also should be due (148 days later).
3. We also know if we have a ram who has a problem performing his assigned duties.

What did we learn this fall? We hoped to breed a total of 63 ewes using a different ram in each of four groups.

1. Of the 63, only one was never marked. We have to assume that she is either, 1) not pregnant, or 2) she had a secret liaison with a ram before breeding officially began. (Yes, it **does** happen!)

314

2. Of the 62 ewes marked, 43 were only marked once. They should all be due within the first 17 days of lambing (March 10th – 27th).

3. Of the 19 who were remarked only three were remarked a third time. This remarking is not a guarantee that those 16 are pregnant, because we removed them at the very end of the second heat cycle. If they are pregnant, lambing will continue through March 29th. Of the remaining three we will just have to wait until we get closer to their due date and look for outward signs of pregnancy. (Or in a few weeks we can have our vets come to ultrasound them.) If they are pregnant lambing will finish on April 8th.

4. Of the four rams we used, only one was responsible for the three third time breedings. We may have to be concerned with his performance.

There it is in a nutshell. Now "all" that remains for us is to get all the equipment and portable fencing in from the pastures before the weather turns cold and snowy. After that is done, we can spend some time being friendly with our ewes again and not have to keep a watch out for a ram who feels that we are threatening his ewes. In between, I can tend to a very sore ankle!

Unplanned Down Time

In December I often find myself looking out the window at a cold, snow covered landscape. That is not an unusual occurrence if you live in Wisconsin in the middle of winter. However, it is only the middle of December and we have already had two good snow storms, including a true blizzard that dumped fourteen inches of heavy, wet snow upon us to be followed by temperatures down to 2° Fahrenheit (-17° Centigrade). In other words, it is the time of year when one expects to make sure that the sheep are well protected in their winter quarters, that the pipes are not frozen in the barn and that you are able to safely hunker down in the warmth of the house. Hopefully, it should be a time for some serious spinning and weaving, along with all the other inside projects that tend to get put aside in favor of the outdoor projects of warmer weather. Unfortunately, for us that is not the way that life has played itself out this winter.

Fall began quite nicely. We got our ewes separated into breeding groups on schedule. We chose to breed a smaller number of ewes than we have in recent years. As a result we had only four groups out on pasture, each with their own ram. The ewes that we chose not to breed were in a separate group that stayed in the barn overnight, but they were able to go out during the day to a pasture separate from the four breeding groups. Much of the planned breeding period turned out to be wet: abnormally heavy amounts of rain accompanied by strong winds. Luckily, the temperatures were above normal during that time, which meant that we did not have to deal with frozen water lines and buckets in the distant pastures. Despite the fact that the breeding groups were starting to look excessively well washed, they appeared to survive quite well. All but one of the 63 ewes were marked by their rams and very few were remarked after their first heat cycle had passed. We are optimistic that we had a nice, tight breeding and, therefore, will have a nice compact lambing schedule this coming March. By mid-November, we were preparing to declare breeding over and bring the breeding groups back home.

Each morning and evening during breeding, we take a small helping of grain out to each breeding group. It is an easy way to get the ewes and ram all close together so that we can see if there are any new breeding marks or remarks. It also aids us if we need to catch one of them to replace a torn jacket or to change the crayon in the ram's marking harness. With pocket notebook in hand, we are able to determine a very exact breeding date for each ewe. The process of feeding grain in long, extended, homemade tubs set out in the pastures is always a bit frantic. It is one of the times when it would be nice to have a good Border Collie keeping the sheep at bay while we pour grain. We have learned a few tricks, however, and manage pretty well with the distribution of grain. We have also made a point of selecting ewes and, especially, rams who are not overly aggressive toward each other or toward us. Nonetheless, one needs to know where all the sheep are when grain is poured since it can become a bit frantic. On a Friday morning in November, I was feeding grain to the first group when seemingly out of nowhere the ewe standing beside me was shoved into the side of my right leg. Ulmer, the ram in this group had, in his eagerness for grain, just rushed into the ewe and crushed her up against me. The ewe seemed to weather the assault nicely. On the other hand, my ankle folded under me. Based upon the pain I was experiencing, I believed that I either had a broken ankle or a very severe sprain. Appropriately enough it was Friday the thirteenth!

The end result of this experience was that I had suffered a severe high ankle sprain. It has meant wearing a large, cumbersome boot to immobilize my leg from just below the knee down to my toes. I was required wear the boot for eight weeks, the first five weeks of which has also required the use of crutches. Subsequently, I have moved on to a lighter brace for chores and am now actively pursuing physical therapy. Progress is excruciatingly slow. I have chosen to relate this experience, not to elicited any sympathy, but to ponder how much we take for granted in our lives and how we are not prepared to deal with the unexpected (especially in a farming environment). What follows are some rather random thoughts on the issues this experience has raised with us.

The Ram Effect

Although I am ticked off that I am to be hobbled for more than three months, I do not entirely blame Ulmer, the ram. His was not an act of aggression toward me. We believe that our selection process has eliminated as much of the aggressive tendencies of rams toward us. But, having just written this, I will also state that we try to keep an eye on our rams whenever we are in a pen/pasture with them (especially during breeding). Rams are in a sense small bulls. A seemingly friendly bull can suddenly turn aggressive and dangerous. The same applies to rams. The only difference is size. It has always been our policy to never take any visitors into a pen which includes a ram. Our ewes are habituated to frequent visits by

318

strangers. Twenty years of relatively peaceful life with our ewes and rams has perhaps caused us to become less watchful than we should be. We had become complacent and we need to re-awaken our awareness of the potential dangers of any animal or group of animals in our care.

Division of Labor

There are two of us who care for our flock: my wife, Gretchen, and myself. In that regard, aside from shearing and the occasional medical emergency involving a vet, we have become rather self-sufficient over the last twenty years. There are many sheep related tasks that either one of us can perform equally well on our own. However, over the years we have also specialized. On the infrequent occasion when one of us is absent, usually due to travel, the other has managed to cope. But we also make sure that those infrequent travel times occur when one of our specialized skills is not required. For example, Gretchen, the former nurse, also has smaller hands. She has thus become the labor and delivery specialist when needed during lambing (while I am the restrainer). I am the person who operates the tractors and who performs most of the "grunt" work. Thus, I am lucky enough(?) to get to spread manure, cut rake, bale and load hay. I am also better at heights than Gretchen, so she gets to unload the hay bales from the wagon onto the elevator while I stack them up in the mow. In all of these and other activities, we have tended to specialize our skills to the point that if, unexpectedly, one of us is physically unable to perform our jobs the other may not be ready or able to take over fully. Had my injury occurred during haying, we would not have been able to continue without finding outside help (a difficult task in our location).

In our current situation, we have had to modify tasks or let a number of them go. Normally we trim hooves when we bring the sheep in from breeding. That chore will now probably have to wait until after shearing. Winter feeding has changed significantly. Each morning and evening I can no longer throw down the six or so bales from the top of the mow and then carry them down a flight of stairs while Gretchen cleans out the feeders. That task is now solely Gretchen's. We have had to switch roles during hay distribution. I can only pass hay over a fence and Gretchen needs to distribute it to each of 15 feeders while surrounded and harassed by all the ewes. Over the years there are tricks to make these tasks easier that we each have learned subconsciously. It is difficult to suddenly try to translate those skills to someone else.

Outside Help

Catching and restraining a ewe (or especially a ram) becomes more difficult with my limited mobility. Hopefully, I will be better suited for the task in about eight weeks when we will be shearing. We are, at least, fortunate to have a small group of sheep and goat raiser friends who we can call upon to help with the major tasks such as shearing. In turn, we are normally able to reciprocate with similar tasks. Last month, without our friends, Gretchen would not have been able to bring in each of our breeding groups, change jackets and sort the rams back into their bachelor quarters. I am not sure what all of us would do without each other's help, at least as long as most of us have the number of animals that we care for. In this regard, I consider our little community much like the older and larger farming community in which it was expected that families would help other families during such events as plantings and harvests. This type of community now only seems to occur in times of severe stress, e.g. barn fires and their aftermath.

Future Planning

I am not sure where our current experience will lead us in terms of planning for the future. It should be easy to look forward to our limitations as we age and plan accordingly, but I am not completely sure whether we have yet done so. Obviously, one cannot anticipate and plan for every possible health related event. Some of our current difficulties would however have been mitigated a bit had we, many years ago, planned for the potential of one of us being incapacitated for a brief or long term. Modifying our current sheep housing and feeding operation to be more shepherd friendly would have been a lot easier when we started our operation, but it would have required a good deal of foresight and knowledge that we probably lacked at the time. If you are young and making long term plans while you start out with your operation, I believe it behooves you not just to think about how your flock may prosper and grow. Learn as much as you can from others who have already been in a similar situation. It is imperative that you recognize that you will not always be as young and strong as you currently find yourself. Lastly, be thankful for your friends and especially your spouse. It is vital that they know that you cannot live without them.

****

Ironically, my run-in with Ulmer and the resulting time for recovery has resulted in the musings that you have just read. Without the time with which this has left me, I would have never convinced myself to sit down long enough to compile these thoughts.

Thanks for accompanying me on this adventure.

The authors at a younger age

Dick Regnery was born and raised in California. After completing high school, he studied in Odense, Denmark, St. Louis, Missouri and Milwaukee, Wisconsin. In 1965, he met his wife-to-be, Gretchen, on a boat to Europe. After an insignificant 15 year career with the Federal Government, he decided to start over with a life of farming in Door County, Wisconsin. Since 1983, Dick and Gretchen have owned Whitefish Bay Farm, where they raise Corriedale sheep, spin and weave their wool which is marketed through a Gallery at the farm. They also operate a Bed & Breakfast.

Nanoo Nanoo was born and raised at Whitefish Bay Farm. She is the daughter of Mindy and Mercury and is the mother of eight lambs (all quite cute). She is noted for her beautiful fleece and wonderful mothering ability. She is now retired from lambing and has become an accomplished, self-taught writer.

Neither author has ever won any literary awards (nor do they expect to do so). Their ongoing written work can be found on the internet in the Ewe Turn blog (http://whitefishbayfarm.com/eweturn).

LaVergne, TN USA
06 November 2010

203721LV00011B/13/P